ELEMENTARY
NUMBER THEORY

Textbooks in Mathematics

Series editors:
Al Boggess, Kenneth H. Rosen

ELEMENTARY NUMBER THEORY

Gove Effinger

Gary L. Mullen

CRC Press
Taylor & Francis Group
Boca Raton London New York

CRC Press is an imprint of the
Taylor & Francis Group, an **informa** business
A CHAPMAN & HALL BOOK

First edition published 2022
by CRC Press
6000 Broken Sound Parkway NW, Suite 300, Boca Raton, FL 33487-2742

and by CRC Press
2 Park Square, Milton Park, Abingdon, Oxon, OX14 4RN

ISBN: 978-1-032-04418-7 (hbk)
ISBN: 978-1-032-01723-5 (pbk)
ISBN: 978-1-003-19311-1 (ebk)

Typeset in CMR10
by KnowledgeWorks Global Ltd.

DOI: 10.1201/9781003193111

Contents

Preface

This text is intended to be used as an undergraduate introduction to the theory of numbers. The authors have delved into and come to love this area of mathematics for many years, and we hope that this text will inspire students (and instructors) to study, understand, and also come to love this truly beautiful subject.

The text is written in a style which is motivated by the Schaum's Outline Series of undergraduate texts covering various areas of mathematics. Each chapter, after an *Introduction*, develops a new topic clearly broken out in *Sections* which include theoretical material together with numerous examples, each worked out in considerable detail. At the end of each chapter, after a *Summary* of the topic, there are a number of *Solved Problems*, also worked out in detail, followed by a set of *Supplementary Problems*. These latter problems give students a chance to test their own understanding of the material; solutions to some but not all of them complete the chapter.

The first eight chapters discuss some standard material in elementary number theory. These topics include divisibility; primes and unique factorization; congruences; solving congruences; Fermat's theorem, Euler's function, and Euler's theorem; applications to modern cryptography; quadratic residues and quadratic reciprocity; and some number theory functions. In particular, the chapter on applications of number theory to modern cryptography highlights the extent to which mathematical ideas have great importance in this "age of the computer."

The remaining chapters discuss topics which might be considered a bit more advanced: Diophantine equations (including a discussion of "Fermat's Last Theorem") and finite fields (including their application to Latin squares and Sudoku squares). We then close with a chapter on *Open Problems in Number Theory*. We strongly encourage students (and of course instructors) to study this chapter carefully and fully realize that not all mathematical issues and problems have been resolved! There is still much to be learned and many questions to be answered in mathematics in general and number theory in particular.

Finally, we provide two appendices on mathematical induction and on sets of numbers beyond the integers for students who may wish to review these topics.

There are of course numerous fine textbooks about number theory in existence should students wish to continue their study beyond this text or perhaps to get different points of view on this material. We recommend in particular the books by Andrews [1], Grosswald [3], Niven and Zuckerman [10], and Silverman [12].

Gove Effinger would like to thank his coauthor for encouraging him to embark on this very enjoyable project; and he would also like to thank his spouse, best friend and fellow mathematician Alice Dean for her love and support.

Gary Mullen would like to thank the Simons Foundation for their support via grant award number 512459. He would also like to thank his wife Bevie Sue Mullen for her patience and understanding during the writing of this text. Finally, he would like to thank Teeba (pictured below) for so patiently sitting on his lap while he proofread many parts of the text.

Gove Effinger

Gary L. Mullen

May 2021

Chapter 1

Divisibility in the Integers ℤ

1.1 Introduction

Number theory is the study of the properties of the set of all integers, i.e., the "whole numbers." We shall throughout this text denote the set $\{\ldots, -3, -2, -1, 0, 1, 2, 3, \ldots\}$ of integers by the symbol $\mathbb{Z}$. In this chapter, we focus on the fact that when we divide one integer by another, we may or may not get an integer as the answer. For example, when we divide 6 by 3, the answer is an integer, but when we divide 6 by 4, it is not. We first discuss the case when the answer *is* in fact an integer (*divisibility*), and following that turn to the case where it may not be (*The Division Algorithm*). We also introduce the concept of the *greatest common divisor* of two positive integers and a method (*The Euclidean Algorithm*) for computing this important number.

1.2 Divisibility

Definition: Suppose b is an integer and a is a non-zero integer. We say that a **divides** b if there is an *integer* q so that $b = aq$. If there are such integers, we denote the fact that a divides b by using the notation $a|b$.

DOI: 10.1201/9781003193111-1

Be aware that the notation $a|b$ is a *sentence* with the verb being "divides." Contrast this with the notation $\frac{a}{b}$, which is an element of the rational numbers $\mathbb{Q}$ (see Appendix B, not a sentence.

Example 1.1. Clearly $2|8$ since $8 = 2(4)$; $36|108$ since $108 = 36(3)$; $3|(-36)$ since $-36 = 3(-12)$; and for any integer m, $3|(15m + 3)$ since $15m + 3 = 3(5m + 1)$. On the other hand, 3 does not divide 13 as there is no integer q with $13 = 3q$.

In the following lemma, we provide a few basic properties involving the divisibility of integers. (Note: A "lemma" is a "helping theorem," i.e., an often easily proved result which is then used to establish bigger results.)

Lemma 1.1. *Let a, b, c, d be integers with $a > 0$ and $d > 0$.*
   (i) *If $a|b$ and $a|c$, then $a|(b + c)$;*
   (ii) *If $a|b$ and $a|c$, then $a|(b - c)$;*
   (iii) *If $a|b$ and $a|c$, then $a|(mb + nc)$ for any integers m and n;*
   (iv) *If $d|a$ and $a|b$, then $d|b$.*

Proof. To prove Part (i), we may assume that $b = aq$ and $c = as$ where q and s are integers. Then $b + c = aq + as = a(q + s)$ so that a divides $b + c$ (since $q + s$ is an integer). The proof of Part (ii) is similar and hence omitted. Proofs of the remaining parts are left to the reader in Supplementary Problem 1.14. $\square$

1.3   The Division Algorithm

What if a positive integer a does *not* divide an integer b? Here is where the seemingly simple but very important *Division Algorithm* comes into play when doing computations in $\mathbb{Z}$.

Theorem 1.2. (Division Algorithm) *Let a and b be integers with $a > 0$. Then there are integers q and r with $0 \leq r < a$ so that $b = aq + r$.*

This is simply a formal statement of the long division process. The integer b is often called the *dividend*, a the *divisor*, q the *quotient*, and r the *remainder*. The key is that the remainder r must be non-negative and must be less than the divisor a. It is clear that $a|b$ if and only if $r = 0$.

Example 1.2. Given $b = 436$ and $a = 17$, we can compute by long division that $436 = 17(25) + 11$. Note that, as required, the remainder 11 is greater than or equal to 0 and is less than the divisor 17. Given $b = -67$ and $a = 12$, we get $-67 = 12(-6) + 5$, so in this case the quotient -6 is negative, but again the remainder 5 must be non-negative and below the divisor 12.

1.4   Greatest Common Divisors

Definition: Given two positive integers a and b, we define the *greatest common divisor* of a and b to be the largest positive integer that divides them both. This number is denoted by $\mathbf{gcd}(a, b)$. We say that two positive integers are *relatively prime* if their greatest common divisor is 1, i.e., if the only divisor they have in common is 1.

Example 1.3. The gcd of 12 and 27 is 3. We have $\gcd(20, 34) = 2$. Since $\gcd(5, 16) = 1$, 5 and 16 are relatively prime. 35 and 63 are not relatively prime since $\gcd(35, 63) = 7$.

1.5   The Euclidean Algorithm

Given two positive integers a and b, how do we compute their greatest common divisor? If a and b are relatively small, one way (which we shall discuss in some detail in our next chapter) is to factor them both into their representation as a product of powers of prime numbers and observe what factors are in common. However, if at least one of the integers a and b is large, this method can be difficult and relatively inefficient. Euclid, in Book VII of his *Elements*, describes a very efficient procedure for computing greatest common divisors. Before stating his method, let us look at a couple of examples.

Example 1.4. (a) What is $\gcd(420, 378)$? As suggested above, one way is to factor them both: $420 = 2^2 \cdot 3 \cdot 5 \cdot 7$ and $378 = 2 \cdot 3^3 \cdot 7$. These factorizations have in common $2, 3$ and 7, so $\gcd(420, 378) = 2 \cdot 3 \cdot 7 = 42$.

(b) What is $\gcd(858, 1092)$? We could factor again, but it does not look easy. We are seeing that the larger the numbers become, the more difficult the factorization method becomes. Here is a better method.

Theorem 1.3. (Euclidean Algorithm) *Let a and b be positive integers. If a divides b, then the $\gcd(a, b) = a$. Otherwise, repeatedly using the Division Algorithm, there exists a strictly decreasing sequence of positive integers $r_1, \ldots, r_n$ so that*

$$
\begin{aligned}
b &= aq_1 + r_1 \\
a &= r_1 q_2 + r_2 \\
r_1 &= r_2 q_3 + r_3 \\
&\vdots \\
r_{n-2} &= r_{n-1} q_n + r_n \\
r_{n-1} &= r_n q_{n+1} + 0.
\end{aligned}
$$

Then $\gcd(a, b) = r_n$.

At first glance this procedure looks complicated, but all it really says is that in order to calculate the greatest common divisor of two positive integers, we repeatedly apply the Division Algorithm, replacing at each step the previous dividend with the previous divisor and the previous divisor with the previous remainder. In doing this, each divisor is strictly smaller than the previous divisor, and hence each remainder is strictly smaller than the previous remainder. It follows that the sequence $\{r_1, \ldots, r_n\}$ is a strictly decreasing sequence of non-negative integers, and so we must eventually get to a remainder of 0. Finally, the last non-zero remainder r_n is indeed a divisor of a and b, and it is in fact the *greatest* such divisor. For verification of these two last assertions, see Solved Problem 1.12.

Example 1.5. (a) Following up on the previous example, we now use the Euclidean Algorithm to find the greatest common divisor of $a = 378$ and $b = 420$:

$$420 \ = \ 378(1) + 42$$
$$378 \ = \ 42(9) + 0.$$

Since the last non-zero remainder is 42, $\gcd(420, 378) = 42$, as we already computed by factoring.

(b) Again from the previous example, what is $\gcd(858, 1092)$? Applying our algorithm:

$$1092 \ = \ 858(1) + 234$$
$$858 \ = \ 234(3) + 156$$
$$234 \ = \ 156(1) + 78$$
$$156 \ = \ 78(2) + 0.$$

Hence $\gcd(858, 1092) = 78$.

Here is a related application of the Euclidean Algorithm. It is often useful in number theory to write the greatest common divisor of two positive integers a and b as a linear combination of them; i.e., find integers x and y, one positive and one negative, so that $\gcd(a, b) = ax + by$. The Euclidean Algorithm provides a systematic way of finding such a pair (x, y). To illustrate this method we use Part (b) of the previous example. The idea is as follows: working from the bottom, we replace each remainder by its expression just above. We number the steps (omitting the last) and we solve each line for the remainder:

(1) $234 = 1092 - 858(1)$
(2) $156 = 858 - 234(3)$
(3) $78 = 234 - 156(1)$.

In line (3), replace 156 by its expression in line (2) and simplify:

$$78 = 234 - 156(1) = 234 - (858 - 234(3)) = 234(4) - 858.$$

Now in this, replace 234 by its expression in line (1) and simplify:

$$78 = 234(4) - 858 = (1092 - 858(1))(4) - 858 = 1092(4) - 858(5).$$

We have succeeded then in writing the gcd 78 as a linear combination of 1092 and 858; that is, our two integers x and y are 4 and -5 respectfully.

We note that when running the Euclidean Algorithm to find the gcd, we do not actually make use of the specific values of the quotients q_i. However, in moving backwards to express the gcd as a linear combination of the original two numbers, we definitely make use of all but the last of those values.

1.6   Summary

In this chapter we have learned what it means for one integer a to divide another integer b and how to use the Division Algorithm to determine whether a does divide b (if the remainder r is 0) or does not divide b (if the remainder r is greater than 0). We also were introduced to the concept of the greatest common divisor of two positive integers, and we learned an efficient method, the Euclidean Algorithm, to compute this gcd and to write it as a linear combination of the two integers.

As we move forward in our exploration of some ideas in number theory, divisibility will continue to play an important role. In particular, in our next chapter we discuss the central role played by numbers among the positive integers greater than 1 which can only be divided by 1 and themselves. These numbers, called *prime numbers*, form the "building blocks" of the integers $\mathbb{Z}$.

1.7 Solved Problems

Divisibility

1.1. Let m and k be arbitrary positive integers. Label each statement as true or false and support your choice.

(a) $5|635$

(b) $-5|635$

(c) $48|124$

(d) $341|32871$

(e) $5|(15m - 10)$

(f) $m|(-3m)$

(g) $(k + m)|(7k + 14m)$

(h) $k|(-6k^2 - k)$

Solution:

(a) $5|635$ is true since $635 = 5(127)$.

(b) $-5|635$ is true since $635 = -5(-127)$.

(c) $48|124$ is false since since the remainder is 28, not 0.

(d) $341|32871$ is false since the remainder is 125, not 0.

(e) $5|(15m - 10)$ is true since $15m - 10 = 5(3m - 2)$.

(f) $m|(-3m)$ is true since $-3m = m(-3)$.

(g) $(k + m)|(7k + 14m)$ is false since the remainder is $7m$, not 0.

(h) $k|(-6k^2 - k)$ is true since $-6k^2 - k = k(-6k - 1)$.

1.2. Show that, for any positive integer n, 3 divides $4^n - 1$.

Solution:

We do a proof by induction (see Appendix A). The base case is $n = 1$ and here the result is obvious. For the inductive step, suppose the result is true for some n; i.e., suppose 3 divides $4^n - 1$. Then $4^{n+1} - 1 = (4^{n+1} - 4) + 3 = 4(4^n - 1) + 3$, and 3 divides both of these terms, so 3 divides their sum. Hence 3 divides $4^{n+1} - 1$, and we are done.

1.3. Is $n^3 - n$ divisible by 6 for each positive integer n? If so, show it, and if not, find an example where it fails.

Solution:
Looking at examples, the result appears to be true for small n; i.e., $2^3 - 2 = 6$, $3^3 - 3 = 24$, $4^3 - 4 = 60$, $5^3 - 5 = 120$, and so on. But that is not a proof that it holds for *all* n. To get the general result, we observe that $n^3 - n = (n - 1)n(n + 1)$, i.e., it's the product of three consecutive integers. But exactly one of any three consecutive integers must be divisible by 3, and at least one of them must be divisible by 2, so the product of the three is divisible by 6.

Division Algorithm

1.4. (a) Using the Division Algorithm, find the quotient q and the remainder r when $b = 711$ is divided by $a = 23$.
(b) Do the same as in Part (a) for dividing $b = -135$ by $a = 31$.

Solution:
(a) $711 = 23(30) + 21$, i.e., $q = 30$ and $r = 21$.
(b) $-135 = 31(-5) + 20$, i.e., $q = -5$ and $r = 20$.

1.5. In the Division Algorithm, suppose that the dividend $b = 259$, the quotient $q = 12$, and the remainder $r = 7$. What is the divisor a?

Solution:
Since $259 = 12a + 7$, we get $a = 252/12 = 21$.

Greatest Common Divisors

1.6. Find $\gcd(35, 180)$ using factorization into powers of prime numbers.

Solution:
$35 = 5 \cdot 7$, $180 = 2^2 \cdot 3^2 \cdot 5$, so $\gcd(35, 180) = 5$.

1.7. Find $\gcd(224, 468)$ using factorization into powers of prime numbers.

Solution:
$224 = 2^5 \cdot 7$, $468 = 2^2 \cdot 3^2 \cdot 13$, so $\gcd(224, 468) = 2^2 = 4$.

1.8. The *least common multiple* of two positive integers a and b, denoted $\text{lcm}(a, b)$, is the smallest positive integer which both a and b divide. For example, $\text{lcm}(8, 12) = 24$ and $\text{lcm}(8, 9) = 72$.

(a) What is $\text{lcm}(9, 15)$?   (b) What is $\text{lcm}(8, 15)$?

Solution:
(a) One way to compute this is to write down the multiples of one argument until you encounter an integer which is divisible by the other. So, the multiples of 15 are 15, 30, 45, and we can stop because 9 divides 45. Hence the answer is 45.

(b) As in Part (a), the multiples of 15 are $\{15, 30, 45, 60, 75, 90, 105, 120\}$, and we finally come to one which is divisible by 8. Hence the answer is 120. Note that $8 \cdot 15 = 120$.

Euclidean Algorithm

1.9. Find $\gcd(35, 180)$ using the Euclidean Algorithm.

Solution:
$$180 = 35(5) + 5,$$
$$35 = 5(7) + 0,$$
so the answer is 5.

1.10. The numbers 13 and 41 are relatively prime since both are themselves prime. Find integers (x, y) such that $13x + 41y = 1$.

Solution:
By the Euclidean Algorithm, we have $41 = 13(3) + 2$ and $13 = 2(6) + 1$.
Hence $1 = 13 - 2(6) = 13 - (41 - 13(3))(6) = 13(19) + 41(-6)$, so $x = 19$ and $y = -6$.

1.11. (a) Find $\gcd(224, 468)$ using the Euclidean Algorithm.
(b) Find integers (x, y) such that $\gcd(224, 468) = 224x + 468y$.

Solution:
(a)
$$468 = 224(2)+20$$
$$224 = 20(11)+4$$
$$20 = 4(5)+0,$$
so the answer is 4.

(b) $4 = 224 - 20(11) = 224 - (468 - 224(2))(11) = 468(-11) + 224(23)$.

1.12. Complete the proof of the Euclidean Algorithm (Theorem 1.3) by making use of Lemma 1.1 Part (iii):
(a) Prove that r_n is a common divisor of a and b.
(b) Prove that r_n is the *greatest* common divisor of a and b.

Solution:
(a) Start at the bottom. By the last step r_n clearly divides r_{n-1}. Moving up, since r_n divides both itself and r_{n-1}, it must divide r_{n-2}, and so on. We eventually arrive at the fact that r_n divides a, and then that it also divides b.
(b) Start at the top now and suppose d is some common divisor of a and b. By the first step, we must have $d|r_1$. Now move down a line, and we get that $d|r_2$. We eventually arrive at the fact that $d|r_n$, and since d was an arbitrary divisor of a and b, r_n must be the greatest such divisor.

1.8   Supplementary Problems

Divisibility

1.13. Let m and k be arbitrary positive integers. Label each statement as true or false and support your choice.
    (a) $7|644$
    (b) $-7|644$
    (c) $24|148$
    (d) $243|2916$

    (e) $4|(16m - 12)$
    (f) $m|(-7m)$
    (g) $(k + m)|(3k + 6m)$
    (h) $k|(-k^3 + 2k)$

1.14. (a) Prove Part (iii) of Lemma 1.1. (Note: Having proved this, note that Parts (i) and (ii) are special cases of Part (iii).)
(b) Prove Part (iv) of Lemma 1.1.

Division Algorithm

1.15. (a) Using the Division Algorithm, find the quotient q and the remainder r when $b = 487$ is divided by $a = 14$.
(b) Do the same as in Part (a) for dividing $b = -386$ by $a = 27$.

1.16. In the Division Algorithm, suppose that the dividend $b = 486$, the quotient $q = 15$, and the remainder $r = 6$. What is the divisor a?

Greatest Common Divisors

1.17. Find $\gcd(44, 111)$ using factorization into powers of prime numbers.

1.18. For a positive integer a, what are the possibilities for the quantity $\gcd(a+3, a)$? Find specific examples to demonstrate each possibility. Now prove your conjecture. (Hint: Suppose d divides both a and $a + 3$, then by Lemma 1.1 Part (ii)$\cdots$.)

1.19. The sequence of numbers $1, 1, 2, 3, 5, 8, 13, 21, \ldots$ is known as the sequence of **Fibonacci number**. After the first two values, a given number is obtained as the sum of the previous two numbers. We denote this sequence of positive integers by $F_1, F_2, F_3, \ldots$, in honor of Fibonacci who first wrote about these numbers in his book "*Liber Abaci*," which was published in 1202.
(a) Above we have written F_1 through F_8. Write down F_9 through F_{12}.
(b) Prove that for any positive integer $k \geq 1$, $\gcd(F_k, F_{k+1}) = 1$, i.e., prove that any two consecutive Fibonacci numbers are relatively prime. (Hint: Suppose d divides both F_{k+1} and F_k; then by

Lemma 1.1 Part (ii) (or (iii)) it divides their difference, which is what by the definition of these numbers? Hence d must divide F_k and F_{k-1}. Continue this process, concluding finally that we must have $d = 1$.

1.20. (a) What is lcm$(21, 28)$?　　(b) What is lcm$(21, 25)$? (See Solved Problem 1.8.)

1.21. Prove that lcm$(a, b) = \frac{a \cdot b}{\gcd(a,b)}$.

Euclidean Algorithm

1.22. Find $\gcd(44, 111)$ using the Euclidean Algorithm.

1.23. The numbers 23 and 71 are relatively prime since both are themselves prime. Find integers (x, y) such that $23x + 71y = 1$.

1.24. (a) Find $\gcd(381, 3837)$ using the Euclidean Algorithm.
(b) Find integers (x, y) such that $\gcd(381, 3837) = 381x + 3837y$.

1.25. How many steps must the Euclidean Algorithm take to find the gcd of two positive integers a and b? We have seen through examples and problems that it can vary, but what is the "worst case?" *It can be shown that if a is the first divisor, then the total number of steps can be no more than 7 times the number of decimal digits of a.*
(a) Suppose the first divisor a is between a billion and 9 billion (and the dividend b is larger). What is the maximum number of steps to discover $\gcd(a, b)$ by the Euclidean Algorithm?
(b) Returning to the Fibonacci numbers (Supplementary Problem 1.19), we saw that $F_{11} = 89$ and $F_{12} = 144$. According to the above information, what is the maximum number of steps needed to compute $\gcd(F_{11}, F_{12})$ using the Euclidean Algorithm? Now do the actual computation and check the number of steps. (Note: This part illustrates that computing the gcd of two adjacent Fibonacci numbers using the Euclidean Algorithm goes about as slowly as possible.)

Answers to Selected Supplementary Problems

1.13. (a) True since $644 = 7(92)$.   (b) True since $644 = -7(-92)$.
(c) False since the remainder is 4, not 0.   (d) True since $2916 = 243(12)$.
(e) True since $16m - 12 = 4(4m - 3)$.   (f) True since $-7m = m(-7)$.
(g) False since the remainder is $3m$, not 0.   (h) True since $-k^3 + 2k = k(-k^2 + 2)$.

1.15. (a) $487 = 14(34) + 11$, i.e., $q = 34$ and $r = 11$.
(b) $-386 = 27(-15) + 19$, i.e., $q = -15$ and $r = 19$.

1.16. Since $486 = 15a + 6$, we get $a = 480/15 = 32$.

1.17. $44 = 2^2 \cdot 11$, $111 = 3 \cdot 37$, so $\gcd(44, 111) = 1$.

1.19. (a) $F_9 = 34$, $F_{10} = 55$, $F_{11} = 89$, $F_{12} = 144$.

1.20. (a) 84.   (b) 525.

1.22. (a) $44 = (2^2)(11)$ and $111 = (3)(37)$, so $\gcd(44, 111) = 1$.
(b) $111 = 44(2) + 23$, $44 = 23(1) + 21$, $23 = 21(1) + 2$, $21 = 2(10) + 1$.

1.23. By the Euclidean Algorithm, we have $71 = 23(3) + 2$ and $23 = 2(11) + 1$.
Hence $1 = 23 - 2(11) = 23 - (71 - 23(3))(11) = 23(34) + 71(-11)$,
so $x = 34$ and $y = -11$.

1.24. (a) $3837 = 381(10) + 27$, $381 = 27(14) + 3$, $27 = 3(9) + 0$, so $\gcd(381, 3837) = 3$.
(b) $3 = 381 - 27(14)$, $27 = 3837 - 381(10)$,
so $3 = 318 - (3837 - 381(10))(14) = 141(381) - 14(3837)$.

1.25. (a) 70 steps maximum.   (b) 14 steps maximum; 11 steps actual.

Chapter 2

Prime Numbers and Factorization

2.1 Introduction

In this chapter we focus on prime numbers, which are, as we said at the end of Chapter 1, the "building blocks" of the integers $\mathbb{Z}$. The set of prime numbers has fascinated mathematicians throughout history, and there continue to be many unsolved mysteries about this set, some of which we shall discuss later in the chapter.

Definition: A positive integer p is **prime** if p has exactly two distinct positive divisors, namely 1 and p itself. If an integer n is not prime, it is said to be **composite**.

Thus 2, 3, 5, 7, and 11 are the five smallest prime numbers. We note that 2 stands apart from the other primes as the only *even* prime. We note also that with this definition the positive integer 1 is *not* prime since it has only one divisor (itself). Early mathematicians may have viewed 1 as being prime, but once we discuss the theory of unique factorization of integers, we will see why it is important not to count 1 as being prime.

DOI: 10.1201/9781003193111-2

2.2   Identifying Primes

A first question we ask about prime numbers is, How can one test whether a given positive integer n is prime? This, in general, turns out not to be an easy task, especially when n is large.

Example 2.1. (a) Are 4997 and 4999 prime? Our only approach would seem to be to try to find relatively small prime divisors of them to establish that they are *not* prime (i.e., are composite). It turns out that 4997 is divisible by 19 (but no smaller prime) and hence is composite, but 4999 is indeed prime.
(b) The number 5357 is not prime since it is divisible by 11. See Problem 2.6 for an easy way to solve this.

2.3   Listing Primes: The Sieve of Eratosthenes

A second question we now ask is, How can we list all of the prime numbers up to some positive value $n \geq 2$? A method to do this is known as the *Sieve of Eratosthenes*, named in honor of Eratosthenes (276BC – 194BC), who appears to be the first to make use of this process. The process, described below, is quite efficient as long as n isn't *too* large.

We begin by listing all of the numbers from 2 to n. Then since 2 is prime, we leave it in the list and delete all multiples of 2 (except 2 itself) up to and including n. That knocks out all the even numbers in our list larger than 2. We then leave 3 and delete all larger multiples of 3. The next value not already deleted is 5, so we leave it and delete all multiples of 5. We continue this process with 7 which is yet to be deleted, then 11, etc. The numbers remaining in the list give all primes up to n. A question you might have is, When can we stop this process so that we have indeed listed all the primes up to n? You are asked in Problem 2.2 to show that *we need only process primes which are less than or equal to the square root of n.*

Example 2.2. We illustrate the Sieve of Eratosthenes by finding all primes up to $n = 50$. We begin by listing all of the positive integers from 2 through 50. By what we just stated, we need only process 2, 3, 5, and 7 since $11 > \sqrt{50}$.

$$
\begin{array}{cccccccccc}
 2 & 3 & 4 & 5 & 6 & 7 & 8 & 9 & 10 \\
11 & 12 & 13 & 14 & 15 & 16 & 17 & 18 & 19 & 20 \\
21 & 22 & 23 & 24 & 25 & 26 & 27 & 28 & 29 & 30 \\
31 & 32 & 33 & 34 & 35 & 36 & 37 & 38 & 39 & 40 \\
41 & 42 & 43 & 44 & 45 & 46 & 47 & 48 & 49 & 50
\end{array}
$$

We boldface the number 2 as the first item in this list (since we know it is prime), and then cross out each multiple of 2 that is greater than 2. It is important to note that no actual arithmetic must be done here! We simply start at 2, skip by the amount of 2 (which gets us to the number 4), cross out the 4, then skip by another 2 to get to 6, cross out the 6, and so on. This stage of the process is quite straightforward. This now leaves us with the following table.

$$
\begin{array}{cccccccccc}
 \mathbf{2} & 3 & \cancel{4} & 5 & \cancel{6} & 7 & \cancel{8} & 9 & \cancel{10} \\
11 & \cancel{12} & 13 & \cancel{14} & 15 & \cancel{16} & 17 & \cancel{18} & 19 & \cancel{20} \\
21 & \cancel{22} & 23 & \cancel{24} & 25 & \cancel{26} & 27 & \cancel{28} & 29 & \cancel{30} \\
31 & \cancel{32} & 33 & \cancel{34} & 35 & \cancel{36} & 37 & \cancel{38} & 39 & \cancel{40} \\
41 & \cancel{42} & 43 & \cancel{44} & 45 & \cancel{46} & 47 & \cancel{48} & 49 & \cancel{50}
\end{array}
$$

Once we have traversed the entire list, we then return to the beginning of the list and look for the first number that has not been selected as a prime already and that has not been crossed out. At this stage, that number is 3. We are guaranteed that this is a prime, so we boldface that number and then cross out all multiples of 3 in the list (again by simply skipping by 3 each time and crossing out the corresponding numbers). That leaves us with the following:

	2	**3**	4̶	5	6̶	7	8̶	9̶	1̶0̶
11	1̶2̶	13	1̶4̶	1̶5̶	1̶6̶	17	1̶8̶	19	2̶0̶
2̶1̶	2̶2̶	23	2̶4̶	25	2̶6̶	2̶7̶	2̶8̶	29	3̶0̶
31	3̶2̶	3̶3̶	3̶4̶	35	3̶6̶	37	3̶8̶	3̶9̶	4̶0̶
41	4̶2̶	43	4̶4̶	4̶5̶	4̶6̶	47	4̶8̶	49	5̶0̶

Returning to the beginning of the list, we see that 5 is the first number which is neither boldfaced nor crossed out. We boldface it and then cross out its multiples.

	2	**3**	4̶	**5**	6̶	7	8̶	9̶	1̶0̶
11	1̶2̶	13	1̶4̶	1̶5̶	1̶6̶	17	1̶8̶	19	2̶0̶
2̶1̶	2̶2̶	23	2̶4̶	2̶5̶	2̶6̶	2̶7̶	2̶8̶	29	3̶0̶
31	3̶2̶	3̶3̶	3̶4̶	3̶5̶	3̶6̶	37	3̶8̶	3̶9̶	4̶0̶
41	4̶2̶	43	4̶4̶	4̶5̶	4̶6̶	47	4̶8̶	49	5̶0̶

As stated at the outset, we need now only process 7 to finish the job. Our list then look like this:

	2	**3**	4̶	**5**	6̶	**7**	8̶	9̶	1̶0̶
11	1̶2̶	13	1̶4̶	1̶5̶	1̶6̶	17	1̶8̶	19	2̶0̶
2̶1̶	2̶2̶	23	2̶4̶	2̶5̶	2̶6̶	2̶7̶	2̶8̶	29	3̶0̶
31	3̶2̶	3̶3̶	3̶4̶	3̶5̶	3̶6̶	37	3̶8̶	3̶9̶	4̶0̶
41	4̶2̶	43	4̶4̶	4̶5̶	4̶6̶	47	4̶8̶	4̶9̶	5̶0̶

Thus we find that the list of all primes up to 50 is given by the following:

$$2, 3, 5, 7, 11, 13, 17, 19, 23, 29, 31, 37, 41, 43, 47.$$

In Problem 2.11 below you will be asked to list all of the primes up to 100.

2.4 Unique Factorization of Integers into Primes

We shall now look closely at the idea that the prime numbers are the "building blocks" of the integers. We first need some additional results on divisibility. The first is the following lemma:

Lemma 2.1. *Let a, b, c be positive integers with a and b relatively prime.*

(i) *If $a|bc$ then $a|c$;*

(ii) *If $a|c$ and $b|c$, then $ab|c$.*

Proof.

(i) Since a and b are relatively prime (i.e., $\gcd(a, b) = 1$), we know from Chapter 1 that the Euclidean Algorithm gives us a way to find integers r and s so that $1 = ar + bs$. Multiplying through by c, we get $c = car + cbs$. Since a clearly divides car and divides cbs by hypothesis, it follows from Lemma 1.1 that a divides the sum on the right. Hence a divides c.

(ii) Since a divides c, ab divides cbs. Moreover, since b divides c, ab divides car. Thus ab divides the sum, which is c. $\square$

Our next result is often attributed to Euclid (and is referred to by many as "Euclid's Lemma"). It follows from our lemma above and provides the foundation for the truly important role that the primes play in the algebraic structure of the integers.

Theorem 2.2. (Euclid's Lemma) *If p is a prime and p divides ab, then p divides a or p divides b.*

(Note: In mathematics, it is agreed that "or" is the "inclusive or"; that is, it means "one or the other *or both.*")

Proof. Since p is a prime, we know $\gcd(p, a)$ is either 1 or p. In the latter case p divides a. If $\gcd(p, a) = 1$ then Part (i) of Lemma 2.1 implies that p divides b. $\square$

This says that if p is a prime and it divides the product of two integers, it must divide one or the other (or both). Note that this is not necessarily true of divisors which are not prime. For example, note that 4 divides $2 \cdot 6 = 12$, but 4 divides neither 2 nor 6.

The following corollary generalizes Euclid's Lemma to a product of $r \geq 1$ integers and can be proved using the method called *induction* (see Appendix A).

Corollary 2.3. *If p is a prime and p divides the product $a_1 \cdots a_r$, then p must divide a_i for some $i = 1, \ldots, r$.*

We now come to the central result of this chapter, often referred to as the "Fundamental Theorem of Arithmetic." It states that any positive integer $n \geq 2$ can be written as a product of (not necessarily distinct) primes, and that with a possible re-ordering of those primes, this product is unique. We remark here that this result is one of many results in mathematics which establish *existence and uniqueness* of some object (in this case, of course, that object being a product of primes equaling a given positive integer).

Theorem 2.4. (Unique Factorization of Integers)
 (i) (*Existence*) *Every positive integer $n \geq 2$ may be written as a product of (not necessarily distinct) prime numbers; i.e., n may be written in the form*

$$n = p_1 \cdots p_r$$

where each integer $p_i, i = 1, \ldots, r$, is a prime.
 (ii) (*Uniqueness*) *Moreover, this factorization is unique except for the order of the primes; i.e., if we also have $n = q_1 \cdots q_s$ where each q_i is a prime, then $r = s$ and (if necessary) upon re-ordering, $p_i = q_i, i = 1, \ldots, r$.*

The proofs of both the existence and uniqueness involve a very powerful proof method called *mathematical induction*. See Appendix A for an introduction to this important idea. There we illustrate the method, in part, by proving both parts of this theorem.

We note in the statement of the Unique Factorization Theorem that $n \geq 2$. This theorem provides a good reason why 1 is not considered to be a prime, for if 1 were a prime, then for example $6 = 2 \cdot 3 = 1 \cdot 2 \cdot 3$ would be two different prime factorizations of 6.

It was stated in the Unique Factorization Theorem that the primes in the representation $n = p_1 \cdots p_r$ are not necessarily distinct. We can, however, collect like primes in the factorization of an integer n and thus write $n = p_1^{a_1} \cdots p_t^{a_t}$ where each $p_i, i = 1, \ldots, t$, is a prime with $p_i \neq p_j$ if $i \neq j$ and each exponent $a_i \geq 1$. This form is often called the *canonical factorization* of the positive integer n.

Example 2.3. The canonical factorization of the integer $n = 1,000$ is given by $n = 2^3 5^3$, while the canonical factorization of the integer $n = 3,500$ is $2^2 5^3 7$.

2.5 The Difficulty of Factorization

Now comes a very important point about primes and factorization: Given a positive integer n, the Unique Factorization Theorem tells us about the guaranteed existence and uniqueness of a prime factorization of n, but it tells us *nothing about how to actually find that factorization.* In fact, if n is a very large number, this is a formidable problem, even with the use of a fast modern computer. We will see in Chapter 6, for example, that when the RSA cryptographic system is employed for the secure transmission of information, the security of the system is based upon the difficulty of factoring a very large number.

Example 2.4. Why are the numbers in Example 2.3 above relatively easy to factor? The reason was that they contain some small primes which are easy to check, so we can "whittle away" at the factorization. But what if there are no small primes around to get us started? For example, it is not easy to find the prime factorization of $n = 4307$, which turns out to be 59 times 73.

2.6 Using Factorization to Compute a GCD

In Chapter 1 we introduced a very efficient procedure for computing the greatest common divisor (gcd) of two positive integers, namely the Euclidean Algorithm. This procedure is probably not, however, how you computed a gcd in the past; rather you probably used factorization. We now formalize this simple method under the assumption that we have been able to find the prime factorizations of two positive integers a and b. We wish to emphasize that for relatively small a and b the process of first factoring and then applying this method generally works well; but as a and b get larger, the Euclidean Algorithm is *much* more efficient.

Theorem 2.5. *Let* $a = p_1^{a_1} p_2^{a_2} \cdots p_r^{a_r}$ *and* $b = p_1^{b_1} p_2^{b_2} \cdots p_r^{b_r}$ *be the canonical factorizations of a and b, respectively, (where perhaps some of the exponents are zero in order to allow a common list of primes to be used for a and b). Here, $p_1, p_2, \ldots, p_r$ are distinct primes, $a_1, a_2, \ldots, a_r \geq 0$ and $b_1, b_2, \ldots, b_r \geq 0$. Then*

$$gcd(a, b) = p_1^{min(a_1,b_1)} p_2^{min(a_2,b_2)} \cdots p_r^{min(a_r,b_r)}$$

where $min(x, y)$ is the smaller of the two values x and y.

Example 2.5. (a) Let us use Theorem 2.5 to calculate the gcd of 350 and 450. The canonical factorizations of 350 and 450 are $2^1 5^2 7^1$ and $2^1 3^2 5^2$ respectively. In order to apply the theorem we include each of the primes $2, 3, 5,$ and 7 in both of our factorizations, hence rewriting the factorizations as $350 = 2^1 3^0 5^2 7^1$ and $450 = 2^1 3^2 5^2 7^0$. Now the theorem tells us that $gcd(350, 450) = 2^1 3^0 5^2 7^0$, i.e., the gcd of 350 and 450 is 50.

2.7   Summary

In this chapter we defined and studied prime numbers, i.e., positive integers greater than 1 which can only be divided by 1 and themselves. We discussed how to try to identify primes and how, using the Sieve of Eratosthenes, to make a complete list of the primes up to a given number. We then stated a central fact about the integers: Every positive integer greater than 1 has a unique factorization into prime numbers (this result often being referred to as "The Fundamental Theorem of Arithmetic"). Hence the prime numbers are quite literally the "building blocks" of the integers. We then discussed the difficulty of finding the factorization of numbers, especially if they are large, and we showed how to use factorization to compute the greatest common divisor (gcd) of two given integers. As we move forward in our study of number theory, prime numbers will play a central role in much of what we learn.

There are a number of questions about the prime numbers which have not been answered as of yet. We strongly urge the reader to study this text's final chapter, *Some Open Problems in Number Theory*, either now or a bit later. It is important to understand that mathematics in general and number theory in particular are not "closed books;" rather, much is not known, and the search for answers is a central endeavor of all mathematicians.

2.8 Solved Problems

Identifying and Listing Primes

2.1. Classify each of the following integers as prime or composite by searching for possible prime divisors: (a) 87, (b) 89, (c) 217, (d) 111, (e) 113

Solution:
(a) composite since $87 = 3 \cdot 29$
(b) prime, since it's not divisible by 2, 3, 5, or 7
(c) composite since $217 = 7 \cdot 31$
(d) composite since $111 = 3 \cdot 37$
(e) prime, since it's not divisible by 2, 3, 5, or 7

2.2. In Parts (b) and (e) of the previous problem, we stopped checking for prime divisors when we got to 7. Here is why that was sufficient. Suppose that n is a composite integer and p is a prime which divides n. Prove that if $p > \sqrt{n}$,, then n must be divisible by another prime q which satisfies that $q < \sqrt{n}$.

Solution:
Proof. Since p divides n, we have $n = pm$ for some integer m. Since $p > \sqrt{n}$, we must have that $m < \sqrt{n}$, for if not, i.e., if $m \geq \sqrt{n}$, then we would have $n = pm > \sqrt{n}\sqrt{n} = n$, which is impossible.

Hence if q is a prime divisor of m (including the case $m = q$), n has a prime divisor q with $q < \sqrt{n}$, as desired.

2.3. (a) According to the result in Problem 2.2, to check the primality of a number which is just below 100, what possible prime divisors must you check?
(b) What about a number just below 1,000?
(c) What about a number just below 10,000?

Solution:
(a) Since $\sqrt{100} = 10$, we need only check $\{2, 3, 5, 7\}$.
(b) Since $\sqrt{1,000} \approx 31.6$, we need to check $\{2, 3, 5, 7, 11, 13, 17, 19, 23, 29, 31\}$.
(c) We must check all 25 of the primes below 100.

2.4. Using the result of Problem 2.2, classify as prime or composite:
(a) 221,   (b) 223.

Solution:
(a) Composite since 13 divides it.
(b) Prime since none of 2, 3, 5, 7, 11, 13 divides it.

Unique Factorization into Primes

2.5. Find the canonical factorizations of
(a) 384,   (b) 1,155,   (c) 9,360.

Solution:
(a) Since $384 = 3 \cdot 128$, the answer is $2^7 \cdot 3$.
(b) $3 \cdot 5 \cdot 7 \cdot 11$.
(c) $2^4 \cdot 3^2 \cdot 5 \cdot 13$.

2.6. Some small primes have simple criteria for divisibility by them in terms of the digits of the number being factored. The primes 2 and 5 have obvious criteria, but 3 and 11 also have less obvious but simple criteria:

An integer n is divisible by 3 if and only if the sum of its decimal digits is also divisible by 3; n is divisible by 11 if and only if the alternating sum (i.e., plus, minus, plus, minus, etc.) of its decimal digits is also divisible by 11.

Use these criteria to determine if the following numbers are divisible by 3 only, 11 only, both 3 and 11, or neither 3 nor 11:
(a) 5,412 (b) 5,421 (c) 5,071 (d) 5,701.

Solution:
(a) both (b) 3 only (c) 11 only (d) neither.

2.7. Suppose w, x, y, and z are the decimal digits of the integer n, i.e., $n = 10^3 w + 10^2 x + 10y + z$. Use the fact that $10 = 3^2 + 1$ to establish the criterion for divisibility by 3.

Solution:
We make use of the Binomial Theorem, which says that if a and b are any numbers, then $(a + b)^3 = a^3 + 3a^2 b + 3ab^2 + b^3$ and $(a + b)^2 = a^2 + 2ab + b^2$. Setting $a = 3^2$ and $b = 1$, we get

$$n = 10^3 w + 10^2 x + 10y + z = (3^2 + 1)^3 w + (3^2 + 1)^2 x + (3^2 + 1)y + z$$

$$= (3^6 + 3 \cdot 3^4 + 3 \cdot 3^2 + 1)w + (3^4 + 2 \cdot 3^2 + 1)x + (3^2 + 1)y + z$$

$$= [(3^6 + 3^5 + 3^3)w + (3^4 + 2 \cdot 3^2)x + 3^2 y] + (w + x + y + z).$$

This last expression has the sum of n's digits on the right, and the rest of it (in brackets) is clearly divisible by 3, so we can conclude that n is divisible by 3 if and only if $w + x + y + z$ is divisible by 3.

Using Factorization to Compute a GCD

2.8. Use factorization to compute the greatest common divisor of 260 and 182.

Solution:
Since $260 = 2^2 \cdot 5 \cdot 13$ and $182 = 2 \cdot 7 \cdot 13$, we get $\gcd(260, 182) = 26$.

2.9. Using Theorem 2.5, find the canonical factorization of $\gcd(2^5 \cdot 3^8 \cdot 5, 3^3 \cdot 5^5 \cdot 7^6)$.

Solution:
$3^3 \cdot 5$.

2.9   Supplementary Problems

Identifying and Listing Primes

2.10. Classify each of the following integers as prime or composite by searching for possible prime divisors: (a) 91,   (b) 89,   (c) 103, (d) 207

2.11. (a) According to Problem 2.2, what primes must you sieve by when using the Sieve of Eratosthenes in order to find all of the primes less than 100?
(b) Now use sieving to list all the primes below 100. (Note: You should find 25 of them.)

2.12. Using the result of Problem 2.2, classify as prime or composite:
(a) 389,   (b) 391.

Unique Factorization into Primes

2.13. Find the canonical factorizations of
(a) 936,   (b) 847,   (c) 3,150.

2.14. Suppose that $n = 10^3 w + 10^2 x + 10y + z$; that is, $w, x, y,$ and z are the decimal digits of n. Use the fact that $10 = 11 - 1$ to establish the criterion for divisibility by 11 given in Problem 2.6 (also see Problem 2.7).

Using Factorization to Compute a GCD

2.15. Use factorization to compute the greatest common divisor of 504 and 1,078.

2.16. Using Theorem 2.5, find the canonical factorization of
$\gcd(2^3 \cdot 3^2 \cdot 5^{10}, 3^4 \cdot 5^5 \cdot 7^3)$.

Answers to Selected Supplementary Problems

2.10. (a) $7 \cdot 13$, (b) prime, (c) prime, (d) $9 \cdot 23$

2.11. (a) $\{2, 3, 5, 7\}$
(b) $\{2, 3, 5, 7, 11, 13, 17, 19, 23, 29, 31, 37, 41, 43, 47, 53, 59, 61, 67,$
  $71, 73, 79, 83, 89, 97\}$

2.12. (a) Prime since none of 2 through 19 divide it.
(b) $17 \cdot 23$

2.13. (a) $2^3 \cdot 3^2 \cdot 13$ (b) $7 \cdot 11^2$ (c) $2 \cdot 3^2 \cdot 5^2 \cdot 7$

2.15. 14

2.16. $3^2 \cdot 5^5$

Chapter 3

Congruences and the Sets $\mathbb{Z}_n$

3.1   Introduction

The set $\mathbb{Z}$ of integers, which is the set we are studying in number theory, is of course an infinite set. It turns out, however, that we can form *finite* sets from the integers in which the arithmetic is done somewhat differently. For example, in $\mathbb{Z}$ we know that $2 + 3 = 5$, but in the set we shall label $\mathbb{Z}_5$, whose elements are $\{0, 1, 2, 3, 4\}$, $2 + 3 = 0$. In $\mathbb{Z}$ we know that $4 \cdot 4 = 16$, but in $\mathbb{Z}_5$, $4 \cdot 4 = 1$. The key tool for forming these sets is an important idea defined on the integers which was originally developed by Carl Friedrich Gauss in his book *Disquisitiones Arithmeticae*, published in 1801. This idea, as we shall see in this and subsequent chapters, plays a key role in number theory as well as in computer science in general and in cryptography in particular.

3.2   Definition and Examples of Congruences

Let $n \geq 2$ be a fixed integer. We define two integers a and b to be **congruent modulo** n if n divides the difference $a - b$. We will denote this by writing $a \equiv b \pmod{n}$. We call the integer n

DOI: 10.1201/9781003193111-3

the **modulus** of the congruence. We note that by the definition of
"divides," $a \equiv b \pmod{n}$ means that $a - b = nk$ for some integer
k. Note that we require n to be greater than 1 since if $n = 1$ then
every integer is equivalent to every other integer.

Probably without realizing it, you have already encountered
congruences in everyday life. For example, the U.S. clock system
works modulo 12 whereas the military clock systems work modulo
24. Days of the week are determined modulo 7 because if a given
day is Monday, then seven days later we have another Monday.
Similarly, except for leap years, our yearly calendars work modulo
365. Let's look at some examples.

Example 3.1. (a) We know $27 \equiv 5 \pmod{11}$ since $27 - 5 = 22 =
11(2)$. Note that 27 is also congruent to 5 modulo 2 since, again,
$27 - 5 = 22 = 2(11)$.
(b) One has to be a bit more careful with negative numbers, but the
idea is the same. For example, $4 \equiv -21 \pmod{5}$ since $4 - (-21) =
4 + 21 = 25 = 5(5)$.

An alternative way to determine if two integers a and b are
congruent modulo n is to use the Division Algorithm to divide each
integer by the modulus n and check to see if the two remainders
are the same. In Example 3.1 we noted that $27 \equiv 5 \pmod{11}$.
Dividing 27 by 11 we obtain a remainder of 5, and when dividing
5 by 11, we also obtain the same remainder of 5. Because these two
remainders are the same, we can conclude that $27 \equiv 5 \pmod{11}$.

Example 3.2. Consider the positive integers 235 and 147 with
modulus $n = 11$. Dividing 235 by 11 we obtain a remainder of
4; similarly dividing 147 by 11 we also obtain a remainder of 4.
So 235 and 147 are congruent modulo 11. As a check we can also
calculate $235 - 147 = 88 = 11(8)$ so that $235 \equiv 147 \pmod{11}$.

3.3   The Finite Sets $\mathbb{Z}_n$

The idea of finding the two remainders upon division by the modulus n leads us to an important point which follows directly from

the Division Algorithm (Theorem 1.2): *Every integer is congruent modulo n to exactly one of n's possible remainders.* For each $n > 1$, this finite set of remainders, i.e., the set $\{0, 1, \ldots, n-1\}$, turns out to be very important because we can do arithmetic inside this set provided that we do the arithmetic modulo n. This set is called the ***integers mod*** n and is denoted $\mathbb{Z}_n$. To emphasize, we repeat:

$$\mathbb{Z}_n = \{0, 1, \ldots, n-1\} \text{ with arithmetic done modulo } n.$$

If a is any integer, we shall use the notation $a \pmod{n}$ to denote the unique remainder of a divided by n, which is of course an element of $\mathbb{Z}_n$. This remainder is also referred to as the *least non-negative residue* of a modulo n. We shall refer to this operation as *reduction mod n*. Note that "$a \pmod{n}$" is an object; "$a \equiv b \pmod{n}$" is a statement.

Example 3.3. The least non-negative residue of 27 modulo 5 (which we are denoting as 27 (mod 5)) is 2 because, by the Division Algorithm, $27 = (5)(5) + 2$. Similarly, the least non-negative residue of -27 modulo 5 (i.e., $-27 \pmod 5$) is 3 since $-27 = (-6)(5) + 3$.

3.4   Addition and Multiplication Tables for $\mathbb{Z}_n$

Since the sets $\mathbb{Z}_n$ are finite, we can write down their complete addition and multiplication tables.

Example 3.4. (a) Here are the addition and multiplication tables for $\mathbb{Z}_5$. Remember that any time an operation done in the integers results in an answer which is greater than or equal to the modulus, we "reduce" that answer by dividing by the modulus and taking the remainder.

+	0	1	2	3	4
0	0	1	2	3	4
1	1	2	3	4	0
2	2	3	4	0	1
3	3	4	0	1	2
4	4	0	1	2	3

·	0	1	2	3	4
0	0	0	0	0	0
1	0	1	2	3	4
2	0	2	4	1	3
3	0	3	1	4	2
4	0	4	3	2	1

Note that in both tables (except for the first row and column in the multiplication table) every element appears exactly once in each row and column.

(b) Here are the tables for $\mathbb{Z}_6$. Note that the addition table looks similar to that of $\mathbb{Z}_5$, but the multiplication table definitely does not. Might this difference have to do with 5 being prime but 6 being composite?

+	0	1	2	3	4	5
0	0	1	2	3	4	5
1	1	2	3	4	5	0
2	2	3	4	5	0	1
3	3	4	5	0	1	2
4	4	5	0	1	2	3
5	5	0	1	2	3	4

·	0	1	2	3	4	5
0	0	0	0	0	0	0
1	0	1	2	3	4	5
2	0	2	4	0	2	4
3	0	3	0	3	0	3
4	0	4	2	0	4	2
5	0	5	4	3	2	1

3.5   Properties of Congruences

Returning now to congruences, we state in the following lemma some of their properties which will be needed as we move forward. We prove several of the properties and leave the remaining proofs for you to do in Problem 3.3 and Problem 3.11. The basic proof technique is to translate a statement about congruences into a corresponding equation in the integers and proceed from there.

Lemma 3.1. *Let $n \geq 2$ be a fixed integer. Assume that $a \equiv b \pmod{n}$ and that $c \equiv d \pmod{n}$. Then*
    (i) $a + c \equiv b + d \pmod{n}$;
    (ii) $a - c \equiv b - d \pmod{n}$;
    (iii) $ac \equiv bd \pmod{n}$;

   (iv) *If m is an integer, then* $ma \equiv mb$ (mod n)*;*
   (v) *If d is a divisor of n, then* $a \equiv b$ (mod d).

Proof. We prove Parts (i), (iii), and (v), leaving proofs of the remaining parts to you. From the assumptions of the lemma, we have that $a - b = nk$ and $c - d = nj$ for some integers k and j. To prove Part (i), we calculate

$$(a + c) - (b + d) = (a - b) + (c - d) = nk + nj = n(k + j).$$

Since $k + j$ is an integer, we can conclude that $a + c \equiv b + d$ (mod n).

For Part (iii) we have that

$$ac = (b+nk)(d+nj) = bd+bnj+dnk+n^2kl = bd+n(bj+dk+nkj).$$

Therefore $ac - bd = n(bj + dk + nkj)$ where $bj + dk + nk$ is an integer, and so $ac \equiv bd$ (mod n).

Finally for Part (v) we have that $n = de$ for some integer e. Hence $a - b = nk = dek = d(ek)$ so that $a \equiv b$ (mod d), as desired. $\square$

   We observe that Parts (i), (ii), and (iii) of Lemma 3.1 tell us that congruences and reduction mod n are "compatible" with integer addition, subtraction, and multiplication; that is, one can add, subtract, and multiply congruences and the new congruence will remain true, with the same modulus. So, now we must ask, what about *division*? It turns out that, in the theory of congruences and in the sets $\mathbb{Z}_n$, sometimes one can divide and sometimes one can't! We note the similarity here to the situation in the integers $\mathbb{Z}$, where division may or may not be possible.

Example 3.5. (a) The statement $40 \equiv 30$ (mod 5) is true. If we divide both sides of the congruence by 2 and leave the modulus the same, we get the statement $20 \equiv 15$ (mod 5), which is also true, so in this case division by 2 is okay. However, if we instead divide both sides of $40 \equiv 30$ (mod 5) by 10, we obtain the statement $4 \equiv 3$ (mod 5), which is, of course, not true. So when can one divide both sides of a congruence and arrive at a new (correct) congruence while maintaining the same modulus? This question is answered in the following lemma.

Lemma 3.2. *If* $ac \equiv bc$ (mod n) *and* c *and* n *are relatively prime, then* $a \equiv b$ (mod n).

Proof. By the definition of congruence, we have that $ac - bc = (a - b)c = nk$ for some integer k. Thus n divides $(a - b)c$. Since c and n are relatively prime, by Part (i) of Lemma 2.1, n must divide $a - b$. Thus $a \equiv b$ (mod n). $\square$

Hence in Example 3.5 it "worked" to divide by 2 since 2 and 5 are relatively prime, but it did not work to divide by 10. See Problem 3.5 for a look at the possibility of simplifying a congruence by dividing out a common factor of both sides and the modulus.

3.6   Doing Division in $\mathbb{Z}_n$

To end this chapter, let's take a closer look at division in the sets $\mathbb{Z}_n$. In the set of real numbers (i.e., all decimal numbers), we know we can always divide by any non-zero number and get a real number as the answer. For example, if we divide 5.4 by 2 we get 2.7. We also know, however, that this is not true in the integers $\mathbb{Z}$; for example, 5 divided by 2 is not an integer. Back in the real numbers, another way to describe always being able to divide is to say that every non-zero real number possesses a (unique) *multiplicative inverse*. For example, in the real numbers the multiplicative inverse of 2 is $1/2 = 0.5$. We note then that dividing by a real number is the same thing as *multiplying by its multiplicative inverse*; for example, in the real numbers, 5.4 divided by $2 = (5.4)(0.5) = 2.7$. We also note that in the integers $\mathbb{Z}$, the only elements possessing multiplicative inverses are 1 and -1, so they are the only numbers by which you can always divide in $\mathbb{Z}$. Thinking of division in terms of multiplicative inverses will help us understand what happens in our new sets $\mathbb{Z}_n$.

We ask then, which elements of $\mathbb{Z}_n$ possess multiplicative inverses and which do not? Now Lemma 3.2 tells us the answer in slightly different language: You can always divide by c in $\mathbb{Z}_n$ provided that c and n are relatively prime. Put in our alternative way, c will possess a multiplicative inverse in $\mathbb{Z}_n$ if and only if c and n are relatively prime. We observe that in the real numbers it's easy to write down the inverse of a non-zero element x (it's just $1/x$), but in $\mathbb{Z}_n$ it's not so obvious. We need to look at some examples.

Example 3.6. (a) The multiplicative inverse (which we denote by a^{-1}) of a non-zero element a is the unique element for which $(a^{-1})(a) = 1$. Looking back at the multiplication table of $\mathbb{Z}_6$ in Example 3.4 Part (b), we see that $1 \cdot 1 = 1$ and $5 \cdot 5 = 1$ (i.e., both are their own multiplicative inverses, just like 1 and -1 in $\mathbb{Z}$), but 2, 3 and 4 do not possess multiplicative inverses. This of course makes sense since 1 and 5 are relatively prime to 6, but 2, 3, and 4 are not.

(b) Which elements of $\mathbb{Z}_{10}$ possess multiplicative inverses and what are they? Since 0, 2, 4, 5, 6, and 8 are not relatively prime to 10, they will not possess multiplicative inverses. However, 1, 3, 7, and 9 will have multiplicative inverses. Doing multiplication modulo 10, we see that $3 \cdot 7 = 1$ (so each is the multiplicative inverse of the other), and $9 \cdot 9 = 1$, i.e., 9 is its own multiplicative inverse.

(c) Looking finally at the multiplication table for $\mathbb{Z}_5$ in Example 3.4 Part (a), it should be no surprise that since 5 is prime, every non-zero element possesses a multiplicative inverse, so in $\mathbb{Z}_5$, you can always divide (except, of course, by 0). Specifically, multiplying modulo 5, we have $2 \cdot 3 = 1$ and $4 \cdot 4 = 1$.

Let us formalize what we have observed in the examples of $\mathbb{Z}_6$, $\mathbb{Z}_{10}$, and $\mathbb{Z}_5$.

Lemma 3.3. *If a is an element of $\mathbb{Z}_n$, then a possesses a multiplicative inverse a^{-1} in $\mathbb{Z}_n$ if and only if a and n are relatively prime.*

Proof. First suppose that a is relatively prime to n. By the Euclidean Algorithm (Theorem 1.3) we know that in $\mathbb{Z}$ we can find integers x and y such that $ax + ny = 1$. In $\mathbb{Z}_n$ then, we have $a(x$ $(\bmod n)) = 1$, i.e., $a^{-1} = x \pmod{n}$.

On the other hand, suppose that a is *not* relatively prime to n and suppose that their gcd is $d > 1$ with $dk = n$ and $jd = a$ (so, in particular, $k > 0$). Suppose d has a multiplicative inverse d^{-1} in $\mathbb{Z}_n$; then, doing arithmetic in $\mathbb{Z}_n$, we have

$$k = (1)k = (d^{-1}d)k = d^{-1}(dk) = d^{-1}0 = 0,$$

which is a contradiction. Thus d does not have a multiplicative inverse in $\mathbb{Z}_n$.

Finally then, just suppose that a has a multiplicative inverse a^{-1} in $\mathbb{Z}_n$; then $1 = a^{-1}a = a^{-1}jd$, so $(a^{-1}j) \pmod{n}$ is a multiplicative inverse of d. This again is a contradiction, so we must conclude that a has no multiplicative inverse, and we are done. $\square$

A final question we should ask about multiplicative inverses in $\mathbb{Z}_n$ is whether there is an efficient method for computing them. The answer is yes and is supplied by the Euclidean Algorithm (see Theorem 1.3 together with the discussion preceding Example 3.6). Given relatively prime integers a and n, we showed how to compute integers x and y such that $ax + ny = 1$. This leads us directly to the following useful lemma:

Lemma 3.4. *If a and n are relatively prime and the Euclidean Algorithm implies that $ax + ny = 1$ for integers x and y, then the multiplicative inverse a^{-1} of a in $\mathbb{Z}_n$ is $x \pmod{n}$.*

Proof. The integer equation $ax + ny = 1$ tells us that $ax \equiv 1$ $(\bmod n)$. Hence in $\mathbb{Z}_n$, $a^{-1} = x \pmod{n}$. $\square$

Example 3.7. Let us use the Euclidean Algorithm to compute the multiplicative inverse of 24 in $\mathbb{Z}_{37}$. Running the algorithm forward, we have

$$
\begin{aligned}
37 &= 24 + 13 \\
24 &= 13 + 11 \\
13 &= 11 + 2 \\
11 &= 5(2) + 1 \\
2 &= 2(1) + 0.
\end{aligned}
$$

Now working backwards from the next-to-last line:

$$1 = 11 - 5(2)$$
$$1 = 11 - 5(13 - 11) = 6(11) - 5(13)$$
$$1 = 6(24 - 13) - 5(13) = 6(24) - 11(13)$$
$$1 = 6(24) - 11(37 - 24) = 17(24) - 11(37).$$

By Lemma 3.4, we conclude that the multiplicative inverse of 24 in $\mathbb{Z}_{37}$ is 17.

3.7 Summary

Carl Friedrich Gauss first defined the concept that two integers a and b are "congruent modulo n" ($n > 1$) provided that $b - a$ is divisible by n. It follows by the Division Algorithm that every integer a is congruent modulo n to a number between 0 and $n - 1$, which are the possible remainders upon division by n. We can now define an arithmetic on the set $\{0, 1, \cdots, n - 1\}$ as follows: perform the operation (i.e., add or multiply) as usual in $\mathbb{Z}$, but then use the Division Algorithm to divide the answer by n and take the remainder of this division as the new answer. The Division Algorithm then guarantees that the new answer lies inside our set. We call this set of n elements, together with this "modular arithmetic," $\mathbb{Z}_n$. Because the set is finite, we can write down complete addition and multiplication tables for $\mathbb{Z}_n$. We then showed that modular arithmetic is "compatible" with regular integer arithmetic, and we examined the question, When can we do division in $\mathbb{Z}_n$ and when can we not? That answer turned out to be that we can divide by an element a of $\mathbb{Z}_n$ if and only if a and n are relatively prime. Finally, we showed that if a does possess a multiplicative inverse a^{-1} in $\mathbb{Z}_n$, we can use the Euclidean Algorithm to compute the value of a^{-1}.

As we said at the outset, the finite sets $\mathbb{Z}_n$ play a key role in much of number theory and theoretical computer science. In the following chapters we take a closer look at congruences and the sets $\mathbb{Z}_n$, aiming toward their key roles in modern cryptography.

3.8   Solved Problems

Congruences

3.1. Classify each of the following congruences as true or false. Support your answer.
 (a) $17 \equiv 5 \pmod 9$
 (b) $33 \equiv 0 \pmod{11}$
 (c) $55 \equiv -9 \pmod{16}$
 (d) $283 \equiv 177 \pmod 5$
 (e) $220 \equiv 34 \pmod 6$
 (f) $17 \equiv -35 \pmod 9$
 (g) $5m + 1 \equiv 2m - 1 \pmod m$ for any integer $m \geq 3$
 (h) $3m + 3 \equiv m^2 - 4m + 3 \pmod m$ for any integer $m \geq 2$

Solution:
(a) False since $17 - 5$ is not divisible by 9.
(b) True since $33 - 0$ is divisible by 11.
(c) True since $55 - (-9) = 64$ is divisible by 16.
(d) False since $283 - 177 = 106$ is not divisible by 5.
(e) True since $220 - 34 = 186$ is divisible by 6.
(f) False since $17 - (-35) = 52$ is not divisible by 9.
(g) False since $(5m + 1) - (2m - 1) = 3m + 2$ is not divisible by any $m \geq 3$.
(h) True since $(3m + 3) - (m^2 - 4m + 3) = -m^2 + 7m$ is divisible by all $m \geq 2$.

3.2. For each of the following congruences, fill in the blank with the least non-negative residue (i.e., with the element $a \pmod n$ in $\mathbb{Z}_n$):
 (a) $17 \equiv$ _____ $\pmod 9$
 (b) $21 \equiv$ _____ $\pmod 7$
 (c) $0 \equiv$ _____ $\pmod 9$
 (d) $-25 \equiv$ _____ $\pmod 6$
 (e) $334 \equiv$ _____ $\pmod{55}$
 (f) $220 \equiv$ _____ $\pmod 6$
 (g) $-221 \equiv$ _____ $\pmod{33}$
 (h) $5m - 1 \equiv$ _____ $\pmod m$ for any integer $m \geq 2$

Solution:
(a) 8, (b) 0, (c) 0, (d) 5, (e) 4, (f) 4, (g) 10, (h) $m - 1$

3.3. Prove Part (ii) of Lemma 3.1.

Solution:
From the assumptions of the lemma, we have that $a - b = nk$ and $c - d = nj$ for some integers k and j. To prove Part (ii), we calculate

$$(a - c) - (b - d) = (a - b) + (d - c) = nk + n(-)j = n(k - j).$$

Since $k - j$ is an integer, we can conclude that $a - c \equiv b - d$ (mod n).

3.4. (a) Prove that if $a \equiv b$ (mod n), then $a^k \equiv b^k$ (mod n) for any positive integer k.
(b) Use Part (a) to prove that for all positive integers k, $8^k - 1$ is divisible by 7.

Solution:
(a) We know that $a - b$ is divisible by n. By basic algebra, we have

$$a^k - b^k = (a - b)(a^{k-1} + a^{k-2}b + a^{k-3}b^2 + \cdots + ab^{k-2} + b^{k-1}).$$

Since n divides the right-hand side, it must divide the left-hand side as well.
(b) $8 \equiv 1$ (mod 7). Now apply Part (a).

3.5. (a) Suppose that $a \equiv b$ (mod n) and suppose that d is a common divisor of a, b, and n. Prove that $a/d \equiv b/d$ (mod n/d).
(b) Use Part (a) to "simplify" the congruence $65 \equiv 50$ (mod 15) to one with a smaller modulus.

Solution:
(a) We are given that n divides $a - b$, i.e., $a - b = nk$ for some integer k. Dividing through by d, we get $a/d - b/d = (n/d)k$, i.e., n/d divides $a/d - b/d$, and so $a/d \equiv b/d$ (mod n/d), as desired.
(b) Dividing through by 5, we get $13 \equiv 10$ (mod 3).

The Sets $\mathbb{Z}_n$

3.6. (a) Write down the 7 by 7 multiplication table of the non-zero elements of $\mathbb{Z}_8$.

(b) Using your table, for the four elements of $\mathbb{Z}_8$ which possess multiplicative inverses, write down what they are (e.g., $3^{-1} =?$, etc.).

Solution:

(a)

·	1	2	3	4	5	6	7
1	1	2	3	4	5	6	7
2	2	4	6	0	2	4	6
3	3	6	1	4	7	2	5
4	4	0	4	0	4	0	4
5	5	2	7	4	1	6	3
6	6	4	2	0	6	4	2
7	7	6	5	4	3	2	1

(b) All four elements possessing multiplicative inverses (i.e., $\{1, 3, 5, 7\}$) are their own multiplicative inverses.

3.7. In $\mathbb{Z}_{14}$ there are six elements which are relatively prime to 14 and hence possess multiplicative inverses. For each of these, determine its multiplicative inverse.

Solution:
In $\mathbb{Z}_{14}$, $1 \cdot 1 = 1$, $3 \cdot 5 = 1$, $9 \cdot 11 = 1$, and $13 \cdot 13 = 1$, so $1^{-1} = 1$, $3^{-1} = 5$, $5^{-1} = 3$, $9^{-1} = 11$, $11^{-1} = 9$ and $13^{-1} = 13$.

3.8. Use the Euclidean Algorithm to compute the multiplicative inverse of 16 in $\mathbb{Z}_{23}$. (See Lemma 3.4 and Example 3.7.)

Solution:
Running the Euclidean Algorithm forwards:

$$
\begin{aligned}
23 &= 16 + 7 \\
16 &= 2(7) + 2 \\
7 &= 3(2) + 1 \\
2 &= 2(1) + 0.
\end{aligned}
$$

Now running it backwards from the next to last line:

$$1 = 7 - 3(2)$$
$$1 = 7 - 3(16 - 2(7)) = 7(7) - 3(16)$$
$$1 = 7(23 - 16) - 3(16) = 7(23) - 10(16).$$

Thus in $\mathbb{Z}_{23}$, $16^{-1} = -10 \pmod{23} = 13$.

3.9   Supplementary Problems

Congruences

3.9. Classify each of the following congruences as true or false. Support your answer.

(a) $23 \equiv 5 \pmod 9$

(b) $33 \equiv 10 \pmod{11}$

(c) $43 \equiv -2 \pmod{15}$

(d) $283 \equiv 178 \pmod 5$

(e) $120 \equiv 34 \pmod 6$

(f) $17 \equiv -37 \pmod 9$

(g) $4m + 2 \equiv 3m - 2 \pmod m$ for any integer $m \geq 5$

(h) $3m + 1 \equiv m^2 - m + 1 \pmod m$ for any integer $m \geq 2$

3.10. For each of the following congruences, fill in the blank with the least non-negative residue (i.e., with the element $a \pmod n$ in $\mathbb{Z}_n$):

(a) $19 \equiv$ _____ $\pmod 8$

(b) $56 \equiv$ _____ $\pmod 7$

(c) $9 \equiv$ _____ $\pmod 9$

(d) $-22 \equiv$ _____ $\pmod 6$

(e) $304 \equiv$ _____ $\pmod{45}$

(f) $221 \equiv$ _____ $\pmod 4$

(g) $-221 \equiv$ _____ $\pmod 4$

(h) $m^2 - 2 \equiv$ _____ $\pmod m$ for any integer $m \geq 3$

3.11. Prove Part (iv) of Lemma 3.1.

3.12. (a) For any positive integer k, find the least non-negative residue of 6^k modulo 10.

(b) Do any other numbers a besides 6 in the set $\{1, 2, \ldots, 9\}$ have this property that the numbers a^k have the same least non-negative residue modulo 10 for all k? If so, which ones?

3.13. Prove that for every positive integer k, $4^{2k} - 11^k$ is divisible by 5. (See Problem 3.4.)

The Sets $\mathbb{Z}_n$

3.14. (a) Write down the 8 by 8 multiplication table of the non-zero elements of $\mathbb{Z}_9$.

(b) Using your table, for the six elements of $\mathbb{Z}_9$ which possess multiplicative inverses, write down what they are (e.g., $2^{-1} = ?$, etc.).

3.15. In $\mathbb{Z}_{15}$ there are eight elements which are relatively prime to 15 and hence possess multiplicative inverses. (Note: We shall see in Chapter 5 an easy way to count such elements.) For each of these, determine its multiplicative inverse.

3.16. Prove that for every $n \geq 2$, $n - 1$ is its own multiplicative inverse in $\mathbb{Z}_n$.

3.17. Use the Euclidean Algorithm to compute the multiplicative inverse of 22 in $\mathbb{Z}_{29}$. (See Lemma 3.4 and Example 3.7.)

Answers to Selected Supplementary Problems

3.9. (a) T,   (b) F,   (c) T,   (d) T,   (e) F,   (f) T,   (g) F,   (h) T

3.10. (a) 3,   (b) 0,   (c) 0,   (d) 2,   (e) 34,   (f) 1,   (g) 3,   (h) $m - 2$

3.12. (a) 6   (b) 1 and 5

3.14.

·	1	2	3	4	5	6	7	8
1	1	2	3	4	5	6	7	8
2	2	4	6	8	1	3	5	7
3	3	6	0	3	6	0	2	6
4	4	8	3	7	2	6	1	5
5	5	1	6	2	7	3	8	4
6	6	3	0	6	3	0	6	3
7	7	5	3	1	8	6	4	2
8	8	7	6	5	4	3	2	1

(a) is indicated to the left of the table.

(b) $1^{-1} = 1$, $2^{-1} = 5$, $4^{-1} = 7$, $5^{-1} = 2$, $7^{-1} = 4$, $8^{-1} = 8$.

3.15. $1^{-1} = 1$, $2^{-1} = 8$, $4^{-1} = 4$, $7^{-1} = 13$, $8^{-1} = 2$, $11^{-1} = 11$, $13^{-1} = 7$, $14^{-1} = 14$

3.17.
Running the Euclidean Algorithm forwards:

$$
\begin{aligned}
29 &= 22 + 7 \\
22 &= 3(7) + 1 \\
2 &= 2(1) + 0.
\end{aligned}
$$

Now running it backwards from the next to last line:

$$
\begin{aligned}
1 &= 22 - 3(7) \\
1 &= 22 - 3(29 - 22) = -3(29) + 4(22)
\end{aligned}
$$

Thus in $\mathbb{Z}_{29}$, $22^{-1} = 4$.

Chapter 4

Solving Congruences

4.1 Introduction

We now turn to the matter of attempting to solve congruences which contain an unknown. We focus in this chapter on linear congruences; i.e., congruences of the form $ax \equiv b \pmod{n}$, where $a \neq 0$ and b are fixed integers, n is a fixed integer greater than 1, and x is an unknown integer. In fact, we shall seek solutions x which lie in $\mathbb{Z}_n$ (i.e., which lie in the set $\{0, 1, 2, ..., n-1\}$), so the set of possible solutions is finite to start with. We note then that solving linear *congruences* in the integers $\mathbb{Z}$ is how we solve linear *equations* in $\mathbb{Z}_n$. We shall see that a given linear congruence may have no solutions, a unique solution, or multiple solutions lying in $\mathbb{Z}_n$. We remark that when working within the set $\mathbb{Z}_n$, we shall continue to use the congruence symbol $\equiv$ to remind you that all arithmetic is being done modulo n.

4.2 Solving a Single Linear Congruence

We begin with some examples.

Example 4.1. (a) Let's solve the congruence $5x \equiv 6 \pmod 8$. Because there are only 8 numbers in the set $\mathbb{Z}_8 = \{0, 1, 2, 3, 4, 5, 6, 7\}$ which are candidates as solutions, we can just check each one. Doing arithmetic modulo 8 (i.e., working in $\mathbb{Z}_8$), we get $5(0) \equiv 0$,

$5(1) \equiv 5$, $5(2) \equiv 2$, $5(3) \equiv 7$, $5(4) \equiv 4$, $5(5) \equiv 1$, $5(6) \equiv 6$, and $5(7) \equiv 3$. Hence we have a unique solution $x = 6$ in $\mathbb{Z}_8$.

(b) Now consider the congruence $6x \equiv 5 \pmod 8$. Again simply checking cases modulo 8, we get $6(0) \equiv 0$, $6(1) \equiv 6$, $6(2) \equiv 4$, $6(3) \equiv 2$, $6(4) \equiv 0$, $6(5) \equiv 6$, $6(6) \equiv 4$, and $6(7) \equiv 2$. We see then that the congruence has no solutions in $\mathbb{Z}_8$.

(c) Finally consider the congruence $6x \equiv 4 \pmod 8$. The cases are the same as in Part (b), so we obtain two solutions in $\mathbb{Z}_8$, $x = 2$ and $x = 6$.

As Theorem 4.1 below proves, the key to the existence and uniqueness of solutions to the linear congruence $ax \equiv b \pmod{}$ is the greatest common divisor d of the coefficient a and modulus n and then d's relationship to b. Looking back at Example 4.1, in Part (a) $\gcd(5,8) = 1$, 1 divides 6, and there is exactly one solution; in Part (b) $\gcd(6,8) = 2$, but 2 does not divide 5, and there are no solutions; and finally in Part (c) $\gcd(6,8) = 2$, 2 *does* divide 4, and we get exactly two solutions. Here is the result we seek.

Theorem 4.1. (i) *The linear congruence $ax \equiv b \pmod n$ has solutions if and only if the greatest common divisor d of a and n divides b.*

(ii) *If d divides b, then there are d distinct solutions modulo n. More specifically, if c is any one of the solutions, then the set of all d solutions is*

$$\{c, c + n/d, c + 2(n/d), \ldots, c + (d-1)(n/d)\},$$

with all numbers reduced modulo n.

Proof. For Part (i), we must prove the implications in both directions. First, suppose the congruence does have a solution $x = c$, then $ac - b = nr$ for some integer r, and hence $b = ac - nr$ must be divisible by d since d divides the right-hand side (by its definition and by Lemma 1.1 Part (ii)).

On the other hand, suppose d divides b, so that $b = de$ for some integer e. Since d is the greatest common divisor of a and n, by the Euclidean Algorithm we have $d = ak + nj$ for some integers k

and j. This gives us $b = ake + nje$, so $a(ke) - b = n(-je)$. Finally, this equation in $\mathbb{Z}$ can be rewritten as the congruence

$$a(ke) \equiv b \pmod{n}.$$

This shows that $ke \pmod{n}$ is a solution of our congruence $ax \equiv b \pmod{n}$, and the proof of Part (i) is complete.

For Part (ii), we must again prove two things: first that the numbers in the given list are indeed solutions; second that any solution c_1 other than c must be in our list. So, first, we know that $ac - b = nr$ (for some integer r). Hence for $1 \leq t \leq d - 1$ (and using the fact that both a/d and n/d are integers), we have

$$a(c + t(n/d)) - b = (ac - b) + at(n/d) = nr + at(n/d) = n(r + (a/d)t).$$

The left-hand and right-hand ends here say that $a(c + t(n/d)) \equiv b \pmod{n}$, as desired.

On the other hand, suppose that c_1 is another element in $\mathbb{Z}_n$ (besides c) which is a solution to our congruence. We have then that n divides both $ac - b$ and $ac_1 - b$, so n divides $a(c - c_1)$, i.e., $a(c - c_1) = ns$ for some integer s. Dividing both sides by d, we obtain $(a/d)(c - c_1) = (n/d)s$, i.e., n/d divides $(a/d)(c - c_1)$. But now by the definition of d, a/d and n/d are relatively prime, so by Lemma 2.1 Part (i), n/d must divide $c - c_1$. Hence $c_1 = c + t(n/d)$ for some integer t, and reducing modulo n if necessary, we see that c_1 is in our given list. Hence our congruence has exactly d solutions in $\mathbb{Z}_n$, and we have identified all of them. $\square$

This result together with other results from our previous three chapters gives us a procedure for solving a given linear congruence. However, if one or more solutions exist, there will always be some work to do to find a first solution. A procedure for solving $ax \equiv b \pmod{n}$ using as small numbers as possible is as follows:

(1) If either a or b is not in $\mathbb{Z}_n$ (i.e., is not in the set $\{0, 1, 2, \ldots, n - 1\}$, reduce it modulo n. We'll continue to label these possibly reduced values in $\mathbb{Z}_n$ as a and b.

(2) Compute $d = \gcd(a, n)$. If d does not divide b, there are no solutions and we are done.

(3) If d *does* divide b, as Problem 3.5 showed us, we can divide our entire congruence through by d, resulting in a new (if $d > 1$) reduced congruence

$$(a/d)x \equiv b/d \pmod{n/d}.$$

(4) This new congruence has a unique solution c in $\mathbb{Z}_{n/d}$ (since a/d and n/d are relatively prime) which we must find by inspection or by using the Euclidean Algorithm to compute the multiplicative inverse $(a/d)^{-1}$ of a/d in $\mathbb{Z}_{n/d}$ (see Lemma 3.4) and then multiplying both sides of the congruence by $(a/d)^{-1}$. That is, we have

$$(a/d)^{-1}(a/d)x \equiv (a/d)^{-1}(b/d) \pmod{n/d},$$

i.e., $x \equiv (a/d)^{-1}(b/d) \pmod{n/d}$, and so our unique solution c is

$$(a/d)^{-1}(b/d) \pmod{n/d}.$$

(See Example 4.4 below for an illustration of this "multiplicative inverse" method.)

(5) Finally, we know from Part (ii) of Theorem 4.1 that the full set of d solutions in $\mathbb{Z}_n$ to our original congruence is

$$\{c, c + n/d, c + 2(n/d), \ldots, c + (d-1)(n/d)\};$$

that is, given one solution, the remaining $d-1$ solutions are spaced evenly across $\mathbb{Z}_n$, n/d units apart.

The following examples illustrate these procedures.

Example 4.2. Consider the congruence $12x \equiv 22 \pmod 8$. We first reduce both 12 and 22 by 8, obtaining the simpler (but equivalent) congruence $4x \equiv 6 \pmod 8$. Since $\gcd(4, 8) = 4$ and since 4 does not divide 6, we have no solutions in $\mathbb{Z}_8$. Also note that if we had not reduced at the start, we have $\gcd(8, 12) = 4$ and 4 does not divide 22, so again we see that there are no solutions.

Example 4.3. Consider the congruence $33x \equiv 15 \pmod{12}$. We first reduce 33 and 15 modulo 12, obtaining $9x \equiv 3 \pmod{12}$. Since $\gcd(9, 12) = 3$ and 3 divides itself, we can divide our congruence through by 3, obtaining

$$3x \equiv 1 \pmod{4}.$$

Now we seek the unique solution c to this new congruence "by hand," since we only have 4 possibilities, and we find that $c = 3$. Finally, the full set of three solutions is $\{3, 3+4, 3+8\} = \{3, 7, 11\}$. For a double-check, we can verify these solutions in the original congruence: $33(3) - 15 = 99 - 15 = 84$, which is divisible by 12; $33(7) - 15 = 231 - 15 = 216$, which is divisible by 12; $33(11) - 15 = 363 - 15 = 348$, which is divisible by 12 as well.

Example 4.4. If now you are thinking that using this procedure turns solving every congruence into an easy task, consider the following example:

$$24x \equiv 5 \pmod{37}.$$

Since 24 and 37 are relatively prime, we know there is a unique solution, but no reductions are possible, so we can either work though the numbers 1 through 36 until we find the one that works, or we can apply the Euclidean Algorithm to find the multiplicative inverse of 24 in $\mathbb{Z}_{37}$. Using this latter method (which will be by far the most efficient if the modulus is large), we solved this exact problem in Example 3.7, getting that $(17)(24) + (-11)(37) = 1$ (check this), and so $(17)(24) \equiv 1 \pmod{37}$, i.e., 17 is the multiplicative inverse of 24 in $\mathbb{Z}_{37}$. Multiplying our congruence through by 17, we obtain $x \equiv (17)(5) \equiv 11 \pmod{37}$, and so our unique solution is 11.

4.3 Solving Systems of Two or More Congruences

Let us now consider the possibility of trying to solve a system of two or more simultaneous linear congruences. Any methods to be discovered here have numerous applications, one of which will appear in Chapter 5 when we learn how to calculate using what's called Euler's function. Here is an example.

Example 4.5. Is there a positive integer x smaller than 77 which has the property that upon division by 7 we get a remainder of 5 and upon division by 11 we get a remainder of 4? If so, is x unique in that range?

How can we go about seeking x? One way is to write down all the numbers between 1 and 76 which are congruent to 4 modulo 11 and then check each of them to see if they are congruent to 5 modulo 7. (We shall call this the "make-a-list" method. We choose to use the modulus 11 first since the list will be shorter than the corresponding list for the modulus 7.) Here is that list: $\{4, 15, 26, 37, 48, 59, 70\}$. Now let's form the corresponding list we get by reducing each of these numbers by 7: $\{4, 1, 5, 2, 6, 3, 0\}$. So the answer to our question is that entry in the former list which corresponds to the 5 (from our modulo 7 congruence) in the latter list, and that is $x = 26$ and is unique in $\mathbb{Z}_{77}$. We notice also that in the list of remainders upon division by 7, every possible remainder value appears exactly once.

Was the fact that each remainder value appears exactly once in this example a coincidence? The answer is no; in fact it occurs because 7 and 11 are relatively prime, as our next important result proves. This theorem appears to have been first published by the Chinese mathematician Sun Tzu, who lived sometime between the third and fifth centuries A.D.

Theorem 4.2. (Chinese Remainder Theorem) *Let $m \geq 2$ and $n \geq 2$ be integers which are relatively prime. Let a and b be integers. Then there is a simultaneous solution to the pair of congruences*

$$x \equiv a \pmod{m}$$

$$x \equiv b \pmod{n}.$$

Moreover this solution is unique modulo mn; i.e., there is only one solution x with $0 \leq x < mn$.

Proof. Existence: Since m and n are relatively prime, the Euclidean Algorithm tells us that there are integers r and s so that $mr + ns = 1$. (Notice how often we use this idea!) We claim that $c = bmr + ans$ is a simultaneous solution to the pair of congruences. We know that

$$c \equiv ans \pmod{m} \text{ and } ns \equiv 1 \pmod{m},$$

and thus $c \equiv a(1) \pmod{m}$. The proof that c is also a solution of the second congruence is similar.

Uniqueness: Assume that c and d are both solutions, then $c \equiv a \pmod{m}$ and $d \equiv a \pmod{m}$. Thus $c - d \equiv 0 \pmod{m}$ and similarly $c - d \equiv 0 \pmod{n}$. Thus $c - d$ is divisible by both m and n, and since m and n are relatively prime, $c - d$ is divisible by the product mn (Lemma 2.1, Part (ii)). Hence $c \equiv d \pmod{mn}$. $\square$

We note that the relative primeness of the moduli is crucial to both parts of the proof. We also note, however, that the theorem gives us little information about how to actually find the simultaneous solution. In Example 4.5 we found the solution by listing all the candidates with respect to one modulus and then testing those with respect to the other. Another approach, which we now illustrate, is to use algebra.

Example 4.6. Consider the two congruences

$$x \equiv 3 \pmod{4}$$
$$x \equiv 4 \pmod{5}.$$

Our method will be to use the definition of congruence to write down an equation (in $\mathbb{Z}$) for x in terms of one modulus and substitute that into the second congruence. So, starting with the larger modulus 5, we have $x - 4 = 5k$, so that $x = 5k + 4$ for some integer k. Substituting this into the other congruence we obtain $5k + 4 \equiv 3 \pmod{4}$, and after reducing the coefficients modulo 4 we have $k \equiv -1 \equiv 3 \pmod{4}$. We choose $k = 3$ since it is in $\mathbb{Z}_4$. Then $x = 5(3) + 4 = 19$. Note that $0 \leq 19 < 4(5) = 20$ so our solution x lies in the correct range. We can easily check that 19 really is the desired solution by plugging it into the two original congruences.

Recall that in high school algebra you probably learned two methods for solving two (or more) simultaneous linear equations in the real numbers: the substitution method and the add or subtract method. Here we are simply using the substitution method adapted to congruences with moduli m and n. For ease of calculations in this setting, it's best to substitute the congruence with the larger modulus into the one with the smaller modulus. Here is another example.

Example 4.7. Let's redo the congruences in Example 4.5 using this algebraic technique.

$$x \equiv 5 \pmod 7$$
$$x \equiv 4 \pmod{11}.$$

From the congruence with the larger modulus we have $x - 4 = 11k$ for some integer k, so $x = 11k + 4$. Substituting this into the other congruence we obtain $11k + 4 \equiv 5 \pmod 7$. After reducing modulo 7 we have that $4k \equiv 1 \pmod 7$, and by testing cases we see that $k = 2$ is a solution to this congruence (we want to select for k an element of $\mathbb{Z}_7$). Hence $x = 11k + 4 = 11(2) + 4 = 26$ is the simultaneous solution to our system of linear congruences, as we already discovered using the make-a-list technique.

It should be no surprise that the conclusions of the Chinese Remainder Theorem do not hold if the hypothesis of relative primeness of the moduli is removed. We illustrate this in the following example, attempting to use the make-a-list technique.

Example 4.8. Consider the following system of congruences:

$$x \equiv 3 \pmod 6$$
$$x \equiv 2 \pmod 8.$$

We make a list of integers up to $(6)(8) = 48$ which are congruent to 2 modulo 8: $\{2, 10, 18, 26, 34, 42\}$. Reducing this list modulo 6, we obtain $\{2, 4, 0, 2, 4, 0\}$. Hence there are no simultaneous solutions to this pair of congruences. Note that if the first congruence had been, say, $x \equiv 4 \pmod 6$, then there would have been solutions (10 and 34), but they obviously would not have been unique in the range $0 \le x < 48$.

Finally, the conclusions of the Chinese Remainder Theorem continue to hold for three or more congruences provided, of course, that all of the moduli are pairwise relatively prime. It is simply a matter of solving two of the congruences, then combining that information with a third congruence, and so on. We illustrate by solving a set of three simultaneous congruences, the latter two being from Examples 4.5 and 4.7.

Example 4.9. Consider the simultaneous congruences

$$x \equiv 2 \pmod{5}$$
$$x \equiv 5 \pmod{7}$$
$$x \equiv 4 \pmod{11}.$$

We know already that for the latter two congruences $x \equiv 26$ (mod 77), so $x = 77j + 26$ for some integer j. Substituting this into the modulo 5 congruence, we get $77j + 26 \equiv 2 \pmod{5}$, and reducing and simplifying we have $2j + 1 \equiv 2 \pmod{5}$, i.e., $2j \equiv 1$ (mod 5). By inspection then, the least non-negative value for j is 3, and we get $x = (77)(3) + 26 = 257$. This value does lie below $(5)(7)(11) = 385$, and you should check that it satisfies all three congruences.

Had we instead used the make-a-list method, the latter list would be
$\{26, 103, 180, 257, 334\}$, and reducing modulo 5 we get $\{1, 3, 0, 2, 4\}$, so our answer is 257.

4.4 Summary

In this chapter we have learned how to solve linear congruences, whether single ones or sets of simultaneous ones. We have emphasized that solving such congruences is essentially solving linear equations in the finite sets $\mathbb{Z}_n$, where n is the modulus of the given congruence. Unlike the case of linear equations in the real numbers, in which there is always a unique solution provided that the coefficient of x is non-zero, linear equations in $\mathbb{Z}_n$ may have a unique solution, or multiple solutions, or possibly no solutions at all. Fortunately, we have specific knowledge, via Theorem 4.1 and the discussion following it, of how to know how many solutions there are and how to find those solutions (if any). Then, using the Chinese Remainder Theorem (Theorem 4.2) and either the make-a-list method or the substitution method, we have a procedure for solving a system of linear congruences provided that the moduli are pairwise relatively prime.

As we move forward, we will need to deal with non-linear congruences, i.e., congruences in which integer exponents > 1 are present. We will consider some cases of this problem in our next chapter, again with an eye toward important applications to cryptography in Chapter 6. See Problem 4.17 for a preview.

4.5   Solved Problems

Solving a Single Congruence

4.1. Determine if each of the following congruences has solutions. If they do, determine the number of non-negative solutions smaller than the modulus n (i.e., lying in $\mathbb{Z}_n$) and then find all such solutions:
  (a) $5x \equiv 6 \pmod{11}$
  (b) $5x \equiv 12 \pmod{20}$
  (c) $8x \equiv 12 \pmod{20}$
  (d) $32x \equiv 48 \pmod{18}$.

Solution:
(a) Since 5 and 11 are relatively prime, there will be a unique solution. Simply checking all the cases, in $\mathbb{Z}_{11}$ we have $5(2) \equiv 10$, $5(3) \equiv 4$, $5(5) \equiv 3$, $5(6) \equiv 8$, $5(7) \equiv 2$, $5(8) \equiv 7$, $5(9) \equiv 1$ and $5(10) \equiv 6$, so $x = 10$.
(b) Since $\gcd(5, 20) = 5$ and since 5 does not divide 12, there are no solutions.
(c) Since $\gcd(8, 20) = 4$ and since 4 divides 12, there will be four solutions spaced $20/4 = 5$ apart in $\mathbb{Z}_{20}$. We need to find one of those solutions, and to get it we can replace our given congruence

with a simpler one by dividing through by 4, obtaining $2x \equiv 3$ (mod 5). Now in $\mathbb{Z}_5$, $2(2) \equiv 4$, $2(3) \equiv 1$, $2(4) \equiv 3$, and so our first solution of the original congruence is $x = 4$. It follows that the four solutions are $\{4, 9, 14, 19\}$. We can easily check this; for example, in $\mathbb{Z}_{20}$, $8(19) = 152 \equiv 12$.

(d) We first reduce modulo 18, so the congruence becomes $14 \equiv 12$ (mod 18). Since $\gcd(14, 18) = 2$ and since 2 divides 12, there will be two solutions in $\mathbb{Z}_{18}$, spaced $18/2 = 9$ apart. We now replace this congruence by dividing through by 2, giving us $7x \equiv 6$ (mod 9) and we seek its unique solution. Making a list, we discover that in $\mathbb{Z}_9$ $7(6) = 42 \equiv 6$, so our solution is $x = 6$. Hence, in $\mathbb{Z}_{18}$, the solutions of the original congruence are $\{6, 15\}$.

4.2. Find all the elements x (if any) of $\mathbb{Z}_{35}$ which have the property that 10 times them is congruent to 5 modulo 35.

Solution:

We wish to solve the congruence $10x \equiv 5$ (mod 35). Since $\gcd(10, 35) = 5$ and since 5 divides 5, we will have five solutions in $\mathbb{Z}_{35}$ spaced $35/5 = 7$ apart. Dividing through by 5, we obtain the simpler congruence $2x \equiv 1$ (mod 7). Since in $\mathbb{Z}_7$, $2(4) \equiv 1$, our unique solution there is $x = 4$. Hence in $\mathbb{Z}_{35}$, the five solutions are $\{4, 11, 18, 25, 32\}$.

4.3. Consider the congruence $17x \equiv 8$ (mod 21). Checking through 21 calculations in $\mathbb{Z}_{21}$ seems inefficient. Instead, use the Euclidean Algorithm (Theorem 1.3) to compute the multiplicative inverse of 17 in $\mathbb{Z}_{21}$ (see Lemma 3.4) and use that to solve the congruence.

Solution:

Running the Euclidean Algorithm forward, we quickly get:

$$21 = 17 + 4$$
$$17 = 4(4) + 1,$$

so working backwards, we get:

$$1 = 17 - 4(4)$$
$$1 = 17 - 4(21 - 17) = 5(17) - 4(21).$$

Hence the multiplicative inverse of 17 in $\mathbb{Z}_{21}$ is 5. Multiplying our given congruence through by 5, we obtain $x \equiv 5(8) \equiv 19$ (mod 21), so $x = 19$. This can be easily checked.

Solving Systems of Congruences

4.4. By the Chinese Remainder Theorem, there is a unique non-negative solution below 77 to the following pair of simultaneous linear congruences. Find it.

$$x \equiv 4 \pmod{7}$$
$$x \equiv 5 \pmod{11}.$$

Solution:
We illustrate both methods we have discussed, first the make-a-list method. The numbers below 77 which are congruent to 5 modulo 11 are $\{5, 16, 27, 38, 49, 60, 71\}$. Reducing this list modulo 7, we get $\{5, 2, 6, 3, 0, 4, 1\}$. Because our modulo 7 congruence contains a 4, we see that the number 60 in the former list corresponds to the 4 in the latter list, so the unique answer in $\mathbb{Z}_{77}$ is 60.

Now employing the substitution method, the second congruence says, by definition, that $x - 5 = 11k$ for some integer k, so $x = 11k + 5$. Substituting this into the first congruence gives us $11k + 5 \equiv 4 \pmod{7}$. Subtracting 5 from both sides and reducing modulo 7, we get $4k \equiv 6 \pmod{7}$, and we are back to solving a single congruence. Since in $\mathbb{Z}_7$, $4(5) = 6$, so $k = 5$, and it follows that $x = 11(5) + 5 = 60$.

4.5. Find the smallest positive solution to the pair of simultaneous linear congruences

$$x \equiv 2 \pmod{6}$$
$$2x \equiv 1 \pmod{7}.$$

Solution:
We start by clearing the coefficient 2 of x in the second congruence by multiplying through by the multiplicative inverse of 2 in $\mathbb{Z}_7$, which is 4. This gives is the equivalent system

$$x \equiv 2 \pmod 6$$
$$x \equiv 4 \pmod 7.$$

Making a list seems simplest here. The numbers below 42 which are congruent to 4 modulo 7 are $\{4, 11, 18, 25, 32, 39\}$. Reducing this list modulo 6 gives us $\{4, 5, 0, 1, 2, 3\}$, so we get $x = 32$. This should be checked in the original system.

4.6. Find the unique non-negative solution below $792 = (8)(9)(11)$ to the simultaneous system of linear congruences

$$x \equiv 5 \pmod 8$$
$$x \equiv 4 \pmod 9$$
$$x \equiv 5 \pmod{11}.$$

Solution:
A unique solution will occur because the three moduli are pairwise relatively prime. We first solve the two latter congruences by the make-a-list method. Numbers below 99 which are congruent to 5 modulo 11 are $\{5, 16, 27, 38, 49, 60, 71, 82, 93\}$. Reducing each of these modulo 9 gives us $\{5, 7, 0, 2, 4, 6, 8, 1, 3\}$, so we get $x \equiv 49$ (mod 99). This says that $x = 99k + 49$ for some integer k, so we plug this into our first congruence, getting $99k + 49 \equiv 5 \pmod 8$, and simplifying and dividing by 8, we get $3k \equiv 4 \pmod 8$. The multiplicative inverse of 3 in $\mathbb{Z}_8$ is 3, and multiplying through we get $k = 3(4) = 4$. Finally then, $x = 99(4) + 49 = 445$. This should be checked in all three original congruences.

4.7. Use the make-a-list method to show that there are no common solutions below 96 to the simultaneous congruences

$$x \equiv 4 \pmod 8$$
$$x \equiv 2 \pmod{12}.$$

Solution:
The list of numbers below 96 which are congruent to 2 modulo 12 is
$\{2, 14, 26, 38, 50, 62, 74, 86\}$. Reducing this list modulo 8, we get $\{2, 6, 2, 6, 2, 6, 2, 6\}$, so there are no solutions to the given system.

4.6   Supplemental Problems

Solving a Single Congruence

4.8. Determine if each of the following congruences has solutions. If they do, determine the number of non-negative solutions smaller than the modulus n (i.e., lying in $\mathbb{Z}_n$) and then find all such solutions:

   (a) $8x \equiv 14 \pmod{36}$
   (b) $8x \equiv 7 \pmod{13}$
   (c) $60x \equiv 24 \pmod{144}$
   (d) $-6x \equiv 48 \pmod{15}$.

4.9. Find all the elements x (if any) in $\mathbb{Z}_{14}$ which have the property that 20 times them is congruent to 10 modulo 14.

4.10. Consider the congruence $11x \equiv 4 \pmod{16}$. Use the Euclidean Algorithm (Theorem 1.3) and Lemma 3.4 to find the multiplicative inverse of 11 in $\mathbb{Z}_{16}$, and use that value to solve the given congruence.

Solving Systems of Congruences

4.11. By the Chinese Remainder Theorem, there is a unique non-negative solution below 63 to the following pair of simultaneous linear congruences. Find it.

$$x \equiv 2 \pmod{7}$$
$$x \equiv 5 \pmod{9}.$$

4.12. Find the smallest positive solution to the pair of simultaneous linear congruences

$$x \equiv 2 \pmod{5}$$
$$4x \equiv 1 \pmod{9}.$$

(Note: See solution of Problem 4.5.)

4.13. By the Chinese Remainder Theorem, there is a unique non-negative solution below 187 to the following pair of simultaneous linear congruences. Find it.

$$x \equiv 5 \pmod{11}$$
$$x \equiv 6 \pmod{17}.$$

4.14. Use the make-a-list method to show that there are no common solutions below 54 to the simultaneous congruences

$$x \equiv 4 \pmod 6$$
$$x \equiv 2 \pmod 9.$$

4.15. The following exercise is Problem 26 in Volume 3 of "Sun Tzu's Mathematical Manual" (circa 300-400 AD): "We have a number of things, but we do not know exactly how many. If we count them by threes, we have one left over. If we count them by fives, we have three left over. If we count them by sevens, we have two left over. How many things are there?" (Note: He no doubt was asking, "What is the smallest possibility for the number of things?" Answer this question.)

4.16. Find the smallest positive integer whose remainder when divided by 11 is 8, which has last digit 4, and is divisible by 27. (Note: This one takes a fair amount of work.)

Looking Ahead

4.17. Here is a preview of some of what we will soon learn about congruences containing exponents:
(a) Compute $2^4 \pmod 5$, $3^4 \pmod 5$ and $4^4 \pmod 5$.
(b) Compute $2^6 \pmod 7$, $3^6 \pmod 7$ and $4^6 \pmod 7$.
(c) Compute $2^{10} \pmod{11}$ and $2^{12} \pmod{13}$.
(d) Using this data, make a guess as to how to complete the following general statement:
      "If p is prime and if a and p are relatively prime, then $\cdots$."

Answers to Selected Supplementary Problems

4.8. (a) no solutions,   (b) 9,
(c) $\{10, 22, 34, 46, 58, 70, 82, 94, 106, 118, 130, 142\}$,   (d) $\{2, 7, 12\}$

4.9. $\{4, 11\}$

4.10. EA forward: $16 = 11 + 5$; $11 = 2(5) + 1$;
EA backwards: $1 = 11 - 2(5) = 11 - 2(16 - 11) = 3(11) - 2(16)$,
so in $\mathbb{Z}_{16}$, $11^{-1} = 3$, and so the solution is $3(4) = 12$.

4.11. 23

4.12. 2

4.13. 159

4.15. 58

4.16. 1944

Chapter 5

The Theorems of Fermat and Euler

5.1 Introduction

In Chapters 3 and 4 we concentrated on congruences which do not contain exponents (besides the "trivial" exponent 1), so now we shall study how to deal with congruences which *do* contain integer exponents greater than 1. It turns out that exponents cannot be dealt with in the same way that numbers in the base can be; but, using results discovered by the great mathematicians Pierre de Fermat (1607 – 1665) and Leonhard Euler (1707 – 1783), we can often simplify exponents which appear in congruences. We shall start with the case of congruences modulo a prime number p (using Fermat's work) and then move to the case of an arbitrary modulus n (using Euler's work). These results, as we shall see in the following chapter, are crucial components of modern cryptography.

5.2 Fermat's Theorem for Prime Moduli

A first immediate question is, Can we reduce an exponent which is larger than a given modulus n in the same way we can reduce

DOI: 10.1201/9781003193111-5

numbers in the base, that is, by dividing by n and taking the remainder? This turns out to be *not true*, and here is a simple example:

Example 5.1. *Question*: What is 2^9 modulo 5?
Correct solution method: $2^9 = 512$, and $512 \pmod 5 = 2$.
Incorrect solution method: Reducing the exponent 9 modulo 5 gives a new exponent of 4. $2^4 = 16$ and $16 \pmod 5 = 1$, which evidently is *wrong*!

Hence we see that in general *we cannot reduce exponents by the modulus* the way we can do with numbers in the base. This is an issue we need to deal with since we could be asked to compute, say, 2^{90} modulo 5. Surely there is a better and faster way to obtain this least non-negative residue than to multiply out the value 2^{90} (which, by the way is a number with 28 decimal digits). Fortunately, there is indeed a reduction method for exponents, which we asked you in Problem 4.17 to form a conjecture about by looking at data. It may help to look back at that problem now.

Example 5.2. Let us take a closer look at one of the pieces of data in Problem 4.17, $2^6 \pmod 7$, and use it to illustrate the proof technique below. Consider the set $\{1, 2, 3, 4, 5, 6\}$ of all non-zero elements of $\mathbb{Z}_7$. If we now multiply each of these elements by the base 2 and then reduce modulo 7, we obtain

$$\{2, 4, 6, 8, 10, 12\} \equiv \{2, 4, 6, 1, 3, 5\} \pmod 7.$$

Note that we get the same set back, but with the numbers rearranged. This says that modulo 7 the products of all the elements of each set will be equal; that is,

$$(1 \cdot 2)(2 \cdot 2)(3 \cdot 2)(4 \cdot 2)(5 \cdot 2)(6 \cdot 2) \equiv (1)(2)(3)(4)(5)(6) \pmod 7,$$

or, gathering the six factors of 2 on the left,

$$(2^6)(1)(2)(3)(4)(5)(6) \equiv (1)(2)(3)(4)(5)(6) \pmod 7.$$

But finally all the numbers $\{1, 2, 3, 4, 5, 6\}$ in the product are relatively prime to 7, so by Lemma 3.2 we may divide them out, leaving us with $2^6 \equiv 1 \pmod 7$.

We arrive then at the following powerful theorem, first discovered by Fermat, in which the one restriction is that *the modulus must be prime*. In the following section we shall be able to remove this restriction.

Theorem 5.1. (Fermat) *If p is a prime and $\gcd(a, p) = 1$, then*

$$a^{p-1} \equiv 1 \pmod{p}.$$

Proof. Consider the product of all the non-zero elements of $\mathbb{Z}_p$, i.e., the product $1(2)(3) \cdots (p-1)$. Also consider the product of each of these elements multiplied by the base element a, i.e., the product

$$(1a)(2a)(3a) \cdots ((p-1)a).$$

We claim that all these multiples of a are distinct modulo p, for if i and j are in $\mathbb{Z}_p$ and $ia \equiv ja \pmod{p}$, then by Lemma 3.2, using the fact that a is relatively prime to p, we can divide by a to obtain $i \equiv j \pmod{p}$; that is, in $\mathbb{Z}_p$, $i = j$. Thus this new set of integers $\{1a, 2a, \ldots, (p-1)a\}$ must be the same modulo p, except for the order, as the set $\{1, 2, \ldots, p-1\}$. It follows that

$$(1a)(2a)(3a) \cdots ((p-1)a) \equiv (1)(2)(3) \cdots (p-1) \pmod{p}.$$

Notice now that we have the common factors $1, 2, \ldots, p-1$ on both sides of the congruence. Each of the values is relatively prime to p, so we can divide each one from both sides of the congruence (again applying Lemma 3.2) to obtain

$$\underbrace{(a)(a) \cdots (a)}_{p-1 \text{ times}} \equiv \underbrace{(1)(1) \cdots (1)}_{p-1 \text{ times}} \equiv 1 \pmod{p}.$$

We have then that $a^{p-1} \equiv 1 \pmod{p}$ and the proof is complete. □

Fermat's Theorem has a nice corollary:

Corollary 5.2. *For any integer a and any prime p, $a^p \equiv a$ (mod p).*

Proof. If a is not divisible by p then Fermat's Theorem implies that $a^{p-1} \equiv 1 \pmod{p}$ and thus by multiplying both sides by a, we have $a^p \equiv a \pmod{p}$. If p divides a, then $a^p \equiv 0 \pmod{p}$ and $a \equiv 0 \pmod{p}$, so $a^p \equiv a \pmod{p}$. $\square$

Let us look at some examples using Fermat's Theorem.

Example 5.3. Because 5 is prime and $\gcd(3,5) = 1$, Fermat's Theorem tells us immediately that $3^4 \equiv 1 \pmod{5}$. Similarly we know that $12^{16} \equiv 1 \pmod{17}$, $54^{96} \equiv 1 \pmod{97}$, and so on. The real power of the theorem is its usefulness whenever the exponent is larger than the modulus p, for then we can reduce the exponent modulo $p-1$. For example, what is 3^{100} modulo 5? By our theorem, we have

$$3^{100} \equiv 3^{4(25)} \equiv (3^4)^{25} \equiv 1^{25} \equiv 1 \pmod{5}.$$

And what if our exponent is not divisible by $p - 1$? Consider the exponent 103 rather than 100. By the Division Algorithm $103 = 4(25) + 3$, and so

$$3^{103} \equiv 3^{4(25)+3} \equiv 3^{4(25)}3^3 \equiv (3^4)^{25}3^3 \equiv 1^{25}3^3 \equiv 3^3 \equiv 2 \pmod{5}.$$

In summary then, Fermat's Theorem tells us that in any congruence with a prime modulus p and a base a which is relatively prime to p, we can reduce any *exponent* modulo $p - 1$ (and of course, we know we can reduce the base a modulo p). Is it possible to get a similar result for congruences involving a modulus n which is not necessarily prime? The answer is yes, as we shall see in the next section.

5.3   Euler's Function and Euler's Theorem

Working about a century after Fermat, Euler saw a way to generalize Fermat's Theorem to congruences involving an arbitrary modulus n. The key idea was to find what quantity with respect to n is the analog of $p-1$ in Fermat's Theorem. To this end, Euler introduced a function which takes in a positive integer n and returns *the number of elements in the range* $\{1, 2, \ldots, n\}$ *which are relatively prime to* n. He denoted this function using the Greek letter phi (ϕ, pronounced "phee"), so it is often referred to as "the Euler ϕ function" as well as "Euler's Function." We note immediately that if p is prime, then $\phi(p) = p - 1$ since *every* non-zero element of $\mathbb{Z}_p$ is relatively prime to p. This function will be the key to generalizing Fermat's Theorem to arbitrary moduli. Let us look at some examples.

Example 5.4. (a) It is easy to check that $\phi(2) = 1$, $\phi(3) = 2$, $\phi(4) = 2$, $\phi(5) = 4$, $\phi(6) = 2$, $\phi(7) = 6$, $\phi(8) = 4$, $\phi(9) = 6$, $\phi(10) = 4$, and so on.

(b) Computing $\phi(20)$, we have that the non-zero elements of $\mathbb{Z}_{20}$ which are relatively prime to 20 are $\{1, 3, 7, 9, 11, 13, 17, 19\}$, so $\phi(20) = 8$. Looking ahead to our next result, let us observe that whereas $\phi(4)\phi(5) = 2 \cdot 4 = 8 = \phi(20)$, $\phi(2)\phi(10) = 1 \cdot 4 = 4 \neq \phi(20)$. This is resulting from the fact that 4 and 5 are relatively prime, but 2 and 10 are not.

(c) Computing $\phi(120)$ "by hand" would be possible but difficult. The next theorem tells us how to break down a "ϕ-calculation" into smaller parts.

Our next theorem tells us how compute Euler's Function for any positive integer n by computing it for each of n's prime-power factors and then multiplying these answers together. This is what we did in Example 5.4, Part (b) by computing $\phi(2^2)$ and $\phi(5)$ to get $\phi(20)$ Again, the key is that 4 and 5 are relatively prime.

Theorem 5.3. (a) *If p is prime and if k is a positive integer exponent, then*

$$\phi(p^k) = p^k - p^{k-1} = p^{k-1}(p-1).$$

(b) *If m_1 and m_2 are relatively prime, then $\phi(m_1 m_2) = \phi(m_1)\phi(m_2)$.*

Example 5.5. Before doing the proof, let's use this result to compute $\phi(120)$, as in Example 5.4 Part (c) above. Since $120 =$

$2^3 \cdot 3 \cdot 5$, we know from Part (b) of Theorem 5.3 that $\phi(120) = \phi(2^3)\phi(3)\phi(5)$. By Part (a) we get that $\phi(2^3) = 2^2(1) = 4$, $\phi(3) = 3^0(2) = 2$, and $\phi(5) = 5^0(4) = 4$. Hence $\phi(120) = 4 \cdot 2 \cdot 4 = 32$.

Proof of Part (a) of Theorem 5.3. There are p^k positive integers a with $1 \le a \le p^k$. Of these values, the ones which are *not* relatively prime to p^k are all the multiples of p, namely

$$1p, 2p, \ldots, pp, (p+1)p, \ldots, p^{k-1}p.$$

There are p^{k-1} values in this list. Hence the remaining values in our set are all relatively prime to p^k, and so $\phi(p^k) = p^k - p^{k-1}$.

We note that for any prime p, we have $\phi(p) = p - 1$.

Example 5.6. Illustrating this proof, consider computing $\phi(5^3)$. The numbers a in the range $1 \le a \le 125$ which are divisible by 5 are

$$\{1 \cdot 5 = 5, 2 \cdot 5 = 10, \ldots, 24 \cdot 5 = 120, 25 \cdot 5 = 125\}.$$

There are $5^2 = 25$ numbers in this list. Hence $\phi(125) = 5^3 - 5^2 = 5^2(5 - 1) = 100$.

Proof of Part (b) of Theorem 5.3. Let S_1 be the set of elements of $\mathbb{Z}_{m_1}$ which are relatively prime to m_1, so the number of elements in S_1 is by definition $\phi(m_1)$. We define S_2 likewise and note that its number of elements is $\phi(m_2)$. Suppose a is in S_1 and b is in S_2, then, since m_1 and m_2 are relatively prime, by the Chinese Remainder Theorem (Theorem 4.2) we know that there is a unique element x in $\mathbb{Z}_{m_1 m_2}$ for which $x \equiv a \pmod{m_1}$ and $x \equiv b \pmod{m_2}$. More-over, x is relatively prime to $m_1 m_2$ since it is relatively prime to both m_1 and m_2. This tells us then that $\phi(m_1 m_2) \ge \phi(m_1)\phi(m_2)$; that is, every element x we have discovered in $\mathbb{Z}_{m_1 m_2}$ is relatively prime to $m_1 m_2$, and there are $\phi(m_1)\phi(m_2)$ such x's, but there may be elements y in $\mathbb{Z}_{m_1 m_2}$ which are themselves relatively prime to $m_1 m_2$ but have not yet been counted. However, if $y \in \mathbb{Z}_{m_1 m_2}$ is relatively prime to $m_1 m_2$, then $a = y \pmod{m_1}$ is relatively prime to m_1 and $b = y$ (mod m_2) is relatively prime to m_2, so y is the solution of the pair of congruences $y \equiv a$ (mod m_1) and $y \equiv b$ (mod m_2); that is, y has indeed been counted already, and we may conclude that $\phi(m_1 m_2) = \phi(m_1)\phi(m_2)$, as desired. $\square$

Example 5.7. Again, an example may help illustrate the argument just given. Let $m_1 = 4$ and $m_2 = 5$, so $S_1 = \{1, 3\}$ and $S_2 = \{1, 2, 3, 4\}$. There are $\phi(4)\phi(5) = 8$ pairs of elements from these two sets. Now, by the Chinese Remainder Theorem the pair $(1, 1)$ gives rise to the element $1 \in \mathbb{Z}_{20}$, and simplifying notation we have $(1, 1) \to 1$, $(1, 2) \to 17$, $(1, 3) \to 13$, $(1, 4) \to 9$, $(3, 1) \to 11$, $(3, 2) \to 7$, $(3, 3) \to 3$, $(3, 4) \to 19$. Hence our list of elements of $\mathbb{Z}_{20}$ arising from this correspondence is $\{1, 17, 13, 9, 11, 7, 3, 19\}$. It is easy to check that this is the complete list of elements of $\mathbb{Z}_{20}$ which are relatively prime to 20, so we see that $\phi(20) = \phi(4)\phi(5) = 2(4) = 8$.

Theorem 5.3 gives us a procedure for computing Euler's Function for an arbitrary positive integer n:
(1) Factor n into its canonical prime power factorization form,
(2) Apply Part (a) of Theorem 5.3 to each prime power factor of n,
(3) Apply Part (b) of Theorem 5.3 by multiplying all the answers in (2) together.

Example 5.8. We make use of our procedure:
(a) $\phi(81) = \phi(3^4) = 3^3(3 - 1) = 27(2) = 54$.

(b) $\phi(490) = \phi(2^1 5^1 7^2) = (2 - 1)(5 - 1)(7^1(7 - 1)) = (1)(4)(42) = 168$.

(c) $\phi(192) = \phi(3^1 2^6) = (3 - 1)(2^5(2 - 1)) = (3)(32) = 96$.

It is important here to emphasize that our ability to use Theorem 5.3 to compute Euler's Function $\phi(n)$ depends first of all on our ability to write n in its canonical factorization. For relatively small n this is usually not too hard to do, but as we get into large numbers, factoring can turn out to be very difficult if not impossible even using today's powerful computers.

Example 5.9. (a) To compute $\phi(3713)$, we must factor 3713. This is not easy "by hand", but it turns out that $3713 = (79)(47)$, both of which are prime, so by our procedure we get $\phi(3713) = \phi(79)\phi(47) = (78)(46) = 3588$.

(b) To compute

$$\phi(14541110479787358873258751014894238824043995436030 07),$$

we must factor this very large number. As in Part (a), it turns out that this number is a product of two primes, each with 25 decimal digits, and it took *Mathematica* 13.6 seconds of CPU on a PC to find these two factors, 5443833602771186901735004 3 and 2671115897514794993926634 9. Had our number been a product of two primes with 30 digits each, *Mathematica* would have had great difficulty factoring it. Had they been 50-digit primes, factorization would have been impossible.

One might well wonder why anyone would want to find the canonical factorization and/or compute the ϕ-function of a 50 or 60 or 100 digit number. It turns out that the great difficulty (or impossibility) of accomplishing these calculations on very large numbers is a cornerstone of the so-called RSA encryption system, which is widely used to ensure secure digital communications. A basic idea of this system is: "Multiplying is easy; factoring is hard." We shall study the RSA system in the following chapter, but now it is time to state Euler's Theorem, a beautiful and important generalization of Fermat's Theorem which turns out to be, among other things, another cornerstone of the RSA system.

Theorem 5.4. (Euler's Theorem) *If a and n are integers with $n > 1$ and with $\gcd(a, n) = 1$, then*

$$a^{\phi(n)} \equiv 1 \pmod{n}.$$

Proof. Because it is a very direct translation of the proof of Fermat's Theorem (Theorem 5.1), we omit the details and recommend looking back at the proof of that theorem. The set $\{1, 2, 3, \ldots, p - 1\}$ there (which has $p - 1$ elements) is now replaced by the set of elements of $\mathbb{Z}_n$ which are relatively prime to n (which set has $\phi(n)$ elements). Now the proof proceeds *exactly* as before. $\square$

Because $\phi(p) = p - 1$ when p is prime, we see that Euler's Theorem is indeed a direct generalization of Fermat's Theorem.

Among other things, Euler's Theorem allows us to reduce exponents in a congruence involving a composite modulus in the same way Fermat's Theorem did for a prime modulus. Let's look at a couple of examples.

Example 5.10. (a) Let us calculate $2^{42} \pmod{75}$. We may apply Euler's Theorem since 2 and 75 are relatively prime. Since $\phi(75) = \phi(25)\phi(3) = (20)(2) = 40$, we have by Euler's Theorem $2^{42} = (2^{40})(2^2) \equiv (1)(4) = 4 \pmod{75}$.
(b) We cannot use Euler's Theorem to compute $5^{42} \pmod{75}$ because 5 and 75 are not relatively prime. If we tried to use it, we would get 25 as the answer, but the correct answer is 50.
(c) We can use Euler's Theorem to compute $3^{98} \pmod{56}$ since 3 and 56 are relatively prime. We have $\phi(56) = \phi(8)\phi(7) = (4)(6) = 24$. Since $98 = 4(24)+2$, we have then $3^{99} = (3^{24})^4(3^2) \equiv (1^4)(3^2) = 9 \pmod{56}$.

5.4 Fast Exponentiation

In trying to do computations with relatively small numbers and in as few steps as possible, these theorems are a great help, but they may not be enough help if the modulus is not small. Let us look at an example now in which we illustrate a method called *fast exponentiation*.

Example 5.11. Suppose that we are asked to compute $3^{99} \pmod{101}$. Since $\phi(101) = 100$ (since 101 is prime), Euler's Theorem cannot be used to reduce the exponent 99. Whatever multiplications we do need to do, we should always remember to *immediately reduce the answer modulo* 101, which will guarantee that all intermediate numbers we arrive at will be smaller than the modulus.

But to get 3^{99}, do we need to do 98 (or so) multiplications? The answer is: far fewer using fast exponentiation. The idea is to do *repeated squaring*, as follows:

(1) Write 99 as a sum of powers of 2: $99 = 64 + 32 + 2 + 1$.

(2) Square 3 and reduce modulo 101: $3^2 \pmod{101} = 9$.

(3) Square $3^2 = 9$ and reduce modulo 101: $3^4 \equiv 9^2 \pmod{101} = 81$.

(4) Square $3^4 = 81$ and reduce modulo 101: $3^8 \equiv 81^2 \pmod{101} = 97$.

(5) Square $3^8 \equiv 97$ and reduce modulo 101: $3^{16} \equiv 97^2 \pmod{101} = 16$.

(6) Square $3^{16} \equiv 16$ and reduce modulo 101: $3^{32} \equiv 16^2 \pmod{101} = 54$.

(7) Square $3^{32} \equiv 54$ and reduce modulo 101: $3^{64} \equiv 54^2 \pmod{101} = 88$.

(8) Finally, by Step 1, $3^{99} = 3^{64}3^{32}3^23^1 \equiv 88 \cdot 54 \cdot 9 \cdot 3 \equiv 34 \pmod{101}$.

Hence in order to do this computation, we required only 8 steps (or, if you consider Step 8 as being three multiplications followed by reductions, 10 steps). This is very efficient, and the beauty of this method is that the larger the modulus, the greater (relatively speaking) the efficiency. For example, suppose we are asked to compute $14^{904,237} \pmod{1,564,379}$ (this last number is prime, but that is not required to apply this method). Since the highest power of 2 below 904,237 is $2^{19} = 524,288$, we'll be able to do this calculation in, say, 20 to 30 steps, and every number we arrive at along the way (after reduction) will be below $1,564,539$. One certainly does not want to do this with a hand calculator, but for a properly programmed computer it's a quite simple. The answer is 601,202, and *Mathematica*© can compute it in less than a tenth of a second. In Problems 5.12 and 5.24 you will be asked to work through two relatively small examples using this method "by hand."

5.5 Summary

In Chapter 3 we learned that in any congruence we can reduce any numbers which are in the base by the modulus (i.e., divide by the modulus and take the remainder), and we then get an equivalent congruence which is easier to work with. In this chapter, however, we learned that we cannot perform this simple reduction on numbers which are exponents. All is not lost, however, because the theorems of Fermat and Euler tell us that *we can indeed reduce exponents as well, but by ϕ of the modulus rather than the modulus itself.* Because ϕ of the modulus is always less than the modulus, we can then always be working with numbers which are less than the modulus. For example, since $\phi(11) = 10$, the congruence $27x^{13} \equiv 46 \pmod{11}$ can be rewritten as the much simpler congruence $5x^3 \equiv 2 \pmod{11}$.

In the following chapter we shall explore some ideas from modern cryptography, many of which depend on:
(1) Reduction of the numbers in congruences (as just discussed) in order to keep computed numbers smaller than the modulus,
(2) Fast exponentiation to compute powers after the exponent has been reduced below ϕ of the modulus, and
(3) the Euclidean Algorithm to compute greatest common divisors and multiplicative inverses (as discussed in Chapters 1 and 3).
Using these tools, very efficient cryptographic methods can be employed, as we shall see shortly.

5.6 Solved Problems

Fermat's Theorem

5.1. In each of the following, use Fermat's Theorem to fill in the blank with the least non-negative residue.
  (a) $3^{15} \equiv$ _____ $\pmod 7$
  (b) $5^{22} \equiv$ _____ $\pmod{19}$
  (c) $3^{144} \equiv$ _____ $\pmod{17}$

Solution:
(a) Since 7 is prime, we can reduce the exponent 15 modulo 6 ($= 7 - 1$), getting a remainder and hence new exponent of 3. Now $3^3 = 27 \equiv 6 \pmod{7}$, so the answer is 6.
(b) We reduce the exponent 22 modulo 18 ($= 19 - 1$), getting a new exponent 4. Now $5^4 = 625 \equiv 17 \pmod{19}$, so the answer is 17.
(c) We reduce 144 modulo 16 ($= 17 - 1$), getting a new exponent 0. Hence the answer is $3^0 = 1$.

5.2. Compute $2^{66} \pmod{29}$.

Solution:
Reducing the exponent 66 modulo 28 ($= 29 - 1$), we get a new exponent 10. Hence $2^{66} \equiv 2^{10} = 1024 \equiv 9 \pmod{29}$.

5.3. Compute $56^{147} \pmod{13}$.

Solution:
Since any number in the base can be reduced modulo 13, we can replace 56 with 4 since $56 = 4(13) + 4$. Now we can reduce the exponent 147 modulo 12 ($= 13 - 1$), getting a new exponent 3. Then $4^3 = 64 \equiv 12 \pmod{13}$.

5.4. Find the least non-negative residue of $2^{10}4^{15}$ modulo 7.

Solution:
We have $2^{10}4^{15} = 2^{10}(2^2)^{15} = 2^{10}2^{30} = 2^{40}$. Reducing the exponent 40 modulo 6 ($= 7 - 1$), we get a new exponent 4. Hence $2^4 = 16 \equiv 2 \pmod{7}$.

Euler's Function and Theorem

5.5. Calculate (a) $\phi(27)$, (b) $\phi(28)$, (c) $\phi(29)$ and (d) $\phi(30)$.

Solution:

(a) We have $27 = 3^3$, so $\phi(27) = 3^2(3-1) = 9(2) = 18$.

(b) We have $28 = 2^2 7^1$, so $\phi = 2(2-1)(7-1) = 2(6) = 12$.

(c) 29 is prime, so $\phi(29) = 28$.

(d) We have $30 = 2^1 3^1 5^1$, so $\phi(30) = (2-1)(3-1)(5-1) = 8$.

5.6. (a) Calculate $\phi(18)$.

(b) Write down the $\phi(18)$ non-zero elements of $\mathbb{Z}_{18}$ which are being counted by ϕ.

Solution:

(a) Since $18 = 2^1 3^2$, we have $\phi(18) = (2-1)3(3-1) = 6$.

(b) $\{1, 5, 7, 11, 13, 17\}$

5.7. Calculate $\phi(525)$.

Solution:

Since $525 = 3^1 5^2 7^1$, we have $\phi(525) = (3-1)5(5-1)(7-1) = 2 \cdot 5 \cdot 4 \cdot 6 = 240$.

5.8. Calculate $\phi(10^8)$.

Solution:

Since $10^8 = 2^8 5^8$, we have $\phi(10^8) = 2^7(2-1)5^7(5-1) = 4(10^7) = 40,000,000$.

5.9. Use Euler's Theorem and the Division Algorithm to compute $2^{187} \pmod{77}$.

Solution:

Since $77 = 7(11)$, we have $\phi(77) = (7-1)(11-1) = 60$. We now reduce the exponent 187 modulo 60, obtaining a new exponent 7. Hence $2^7 = 128 \equiv 51 \pmod{77}$, so the answer is 51.

5.10. (a) Prove that if $n = 2^k$ for some positive integer k, then $\phi(n) = n/2$.

(b) Describe the non-zero elements of $\mathbb{Z}_n$ which are being counted by $\phi(n)$ in this case.

Solution:

(a) For positive k, $\phi(2^k) = 2^{k-1}(2-1) = 2^k 2^{-1} = n/2$.

(b) All odd numbers from 1 through $n-1$.

5.11. (a) Prove that if $\{p_1, p_2, \ldots, p_r\}$ is the set of distinct primes dividing n, then

$$\phi(n) = n(1 - \frac{1}{p_1})(1 - \frac{1}{p_2}) \cdots (1 - \frac{1}{p_r}).$$

(b) Use Part(a) to compute $\phi(3600)$.

Solution:

(a) Suppose $n = p_1^{k_1} p_2^{k_2} \cdots p_r^{k_r}$. For each prime p_i, we know that

$$\phi(p_i^{k_i}) = p_i^{k_i-1}(p_i - 1) = p_i^{k_i} \frac{1}{p_i}(p_i - 1) = p_i^{k_i}(1 - \frac{1}{p_i}).$$

Multiplying all of these r formulas for $\phi(p_i^{k_i})$ together, we obtain the claimed formula for $\phi(n)$.

(b) Since 3600 divisible by 2, 3 and 5 only, from Part (a) we get

$$\phi(3600) = 3600(1 - \frac{1}{2})(1 - \frac{1}{3})(1 - \frac{1}{5}) = 3600(\frac{1}{2})(\frac{2}{3})(\frac{4}{5}) = 960.$$

Note: So as this example illustrates, since only 2, 3 and 5 divide 3600, in computing $\phi(3600)$ we first retain half of the numbers below 3600, then two-thirds of the remaining numbers, and finally four-fifths of the still remaining numbers. The interesting feature of this formula is that you don't have to factor 3600 (or more generally n); you only have to identify all the primes dividing it.

Fast Exponentiation

5.12. Using fast exponentiation by hand and showing all the steps, compute 3^{18} (mod 50). (See Example 5.11.)

Solution:

(1) Write 18 as a sum of powers of 2: $18 = 16 + 2$.

(2) Square 3 and reduce modulo 50: 3^2 (mod 50) $= 9$.

(3) Square $3^2 = 9$ and reduce modulo 50: $3^4 \equiv 9^2 \pmod{50} = 31$.

(4) Square $3^4 = 31$ and reduce modulo 50: $3^8 \equiv 31^2 \pmod{50} = 11$.

(5) Square $3^8 \equiv 11$ and reduce modulo 50: $3^{16} \equiv 11^2 \pmod{50} = 21$.

(6) Finally, by Step 1, $3^{18} = 3^{16}3^2 \equiv 21 \cdot 9 \equiv 39 \pmod{50}$.

5.7 Supplementary Problems

Fermat's Theorem

5.13. In each of the following, use Fermat's Theorem to fill in the blank with the least non-negative residue.
  (a) $2^{10} \equiv$ _____ $\pmod 7$
  (b) $4^{22} \equiv$ _____ $\pmod{19}$
  (c) $5^{131} \equiv$ _____ $\pmod{17}$

5.14. Compute $3^{125} \pmod{31}$.

5.15. Compute $104^{564} \pmod{11}$.

5.16. Calculate $3^{30}7^{20} \pmod 5$.

Euler's Function and Theorem

5.17. Calculate (a) $\phi(64)$, (b) $\phi(65)$, (c) $\phi(66)$ and (d) $\phi(67)$.

5.18. (a) Calculate $\phi(24)$.
(b) Write down the $\phi(24)$ non-zero elements of $\mathbb{Z}_{24}$ which are being counted by ϕ.

5.19. Calculate $\phi(594)$.

5.20. Calculate $\phi(2407)$. (Hint: This number is the product of two primes, one of which is less than 30. This illustrates that even relatively small numbers can be hard to factor.)

5.21. Use Euler's Theorem and the Division Algorithm to compute $2^{152} \pmod{135}$.

5.22. (a) Prove that if n is greater than 2, then $\phi(n)$ must be even.
(b) Describe all integers $n \geq 5$ which have the property that 4 does *not* divide $\phi(n)$. (Note: This requires some careful thought.)

5.23. The corollary of Fermat's Theorem (Corollary 5.2) says that if p is prime and a is *any* integer, then $a^p \equiv a \pmod{p}$. One might conjecture then that an analogous corollary of Euler's Theorem might be true; that is, perhaps for every modulus $n > 1$ and every integer a (whether or not a is relatively prime to n), $a^{\phi(n)+1} \equiv a \pmod{n}$. Find a single example for which this congruence does *not* hold, telling us that the conjecture is not in general true.

Fast Exponentiation

5.24. Using fast exponentiation by hand and showing all the steps, compute $4^{19} \pmod{25}$. (See Example 5.11 or Problem 5.12.)

5.25. Concerning fast exponentiation, one can get a pretty good estimate of the maximum number of steps required by counting the number of decimal digits in the modulus and then multiplying that by twice the base 2 log of 10, which value is a little less than 7. According to this, what is the maximum number of steps required if the modulus is (a) 2,376 (b) 5,768,105,357,451?

Answers to Selected Supplementary Problems

5.13. (a) 2,   (b) 9,   (c) 6.

5.14. 26

5.15. 9

5.16. 4

5.17. (a) 32,   (b) 48,   (c) 20,   (d) 66.

5.18. (a) 8,   (b) $\{1, 5, 7, 11, 13, 17, 19, 23\}$.

5.19. 180

5.20. $2407 = 29 \cdot 83$, so $\phi(2407) = 28 \cdot 82 = 2296$.

5.21. 121

5.24. 19

5.25. (a) 28 at most,   (b) 91 at most.

Chapter 6

Applications to Modern Cryptography

6.1 Introduction

In this chapter, we study some techniques from modern cryptography which make use of many of the ideas introduced in the previous three chapters; that is, congruences, Fermat's Theorem, Euler's Function and Euler's Theorem. This should indicate clearly that number theory has a profound influence on our "computer age," though this influence is usually unseen by the average computer user. By *cryptography*, we mean the science of secure communication. The term *encryption* refers to the actual implementation of some form of cryptography.

The reason that congruences play such an important role in cryptography, and in computer programming in general, is that using congruences guarantees that the programs will not have to deal with arbitrarily large numbers. Once a modulus is set, then the results of all calculations done are immediately reduced by the modulus, and so each outcome is smaller than the modulus. We shall see various examples of this as we move forward.

DOI: 10.1201/9781003193111-6

6.2 The Basics of Encryption

Let us start by discussing the basic idea of encryption and the need for one or more *keys* for that encryption to work. Here is an example of a relatively simple encryption technique.

Example 6.1. Suppose I want to send the message "SELL" to my stock broker but I want the message to be encrypted so that only she knows what I want. A simple technique is called *linear encryption*, which in this case would involve using a modulus of 26 and two keys, a multiplier m (which must be relatively prime to 26) and an adder b. Then if x is a numerical representation of a letter (say A = 0, B = 1, and so on), then my encryption of each letter x will be $y = mx + b \pmod{26}$ and her decryption would then be $x = m^{-1}(y-b) \pmod{26}$. What she and I need to do before any communication can occur is *agree on our two private keys* m and b, and, using the Euclidean Algorithm, we can each compute the multiplicative inverse m^{-1} of m in $\mathbb{Z}_{26}$. So, suppose we agree on $m = 5$ and $b = 12$. Since S = 18, E = 4 and L = 11, I now do the three computations modulo 26: $5(18) + 12 \equiv 24$, $5(4) + 12 \equiv 6$, and $5(11) + 12 \equiv 15 \pmod{26}$, and send over to her $\{24, 6, 15, 15\}$ (that is, "YGPP"). Since $m^{-1} = 21$, she now computes (again modulo 26) $21(24 - 12) \equiv 18$, $21(6 - 12) \equiv 4$, and $21(15 - 12) \equiv 11$; that is, she decrypts the message to $\{18, 4, 11, 11\}$, which is, of course, "SELL."

The primary point for us in this example is that my broker and I need to share a private key (or keys) to do secure communication, but how can we *securely* agree on the shared keys? The point is that we need shared secure keys to communicate, but, ironically, we need other keys to communicate our desired keys, and so on. What was traditionally used was a "trusted carrier," but who/what is that? An answer to this seemingly unsolvable situation would be some method by which a key or keys can be agreed upon with an unsecured communication exchange which does not reveal what the keys are to anyone but the two communicators, even if the process is somehow hacked by an outsider. This is what Diffie and Hellman devised in the 1970's (see [2]). Their system, which we describe below, was for example used during the Cold War when the United States and the Soviet Union wanted to establish their "hotline," and it continues to be used in setting up extremely fast and secure modern computer communication channels.

6.3 Primitive Roots in $\mathbb{Z}_p$

Before describing the Diffie-Hellman method, we need to investigate the multiplicative structure of $\mathbb{Z}_p$ where p is prime. Fermat's Theorem tells us that if a is a non-zero element of $\mathbb{Z}_p$, the integers modulo p, then $a^{p-1} \equiv 1 \pmod{p}$. A question is: Is $p-1$ the *smallest* exponent on a for which the reduced answer is 1? Let's look at a couple of examples.

Example 6.2. (a) For each non-zero element a of $\mathbb{Z}_{11}$, the following chart shows the smallest power k for which $a^k \equiv 1 \pmod{11}$:

a	1	2	3	4	5	6	7	8	9	10
k	1	10	5	5	5	10	10	10	5	2.

We see then that four of the ten non-zero elements (2, 6, 7 and 8) of $\mathbb{Z}_{11}$ need to be raised all the way to the 10-th power to get to 1. Such an element will be called a *primitive root* of $\mathbb{Z}_{11}$ (see the general definition below). We also note that the number of primitive roots here, namely 4, is given by $\phi(11-1) = 4$ (ϕ of course being Euler's Function). This will in fact be true in general: the number of primitive roots of the prime p is $\phi(p-1)$. Hence we have a way of counting how many primitive roots there are in $\mathbb{Z}_p$, but not necessarily an easy way to identify which elements they are.

(b) If we were to do the same experiment for $p = 7$, we would find that there are two primitive roots (namely 3 and 5), i.e., two elements which must be raised all the way to the 6-th power modulo 7 to get an answer of 1. Note that $\phi(7-1) = 2$.

We make then the following definition: Suppose a is a non-zero element of $\mathbb{Z}_p$ where p is prime and suppose that k is the smallest exponent such that $a^k \equiv 1 \pmod{p}$. If $k = p - 1$, then a is called a *primitive root* of $\mathbb{Z}_p$.

6.4   Diffie-Hellman Key Exchange

We are now in a position to describe Diffie's and Hellman's idea. Let us suppose that Alice and Bob wish to communicate securely by setting up a common key only they will know. This common, secure key might be used, for example, to be the multiplier m in a linear encryption system, as discussed previously. Here are the steps they follow:

(1) They agree upon a large prime p to act as the modulus (in practice, p may have 100 decimal digits or more!), and they agree upon a primitive root g of $\mathbb{Z}_p$. We leave aside the difficulties involved in identifying such a large prime and such a primitive root, but there are quick algorithms to accomplish both of these tasks.

(2) Alice chooses a number a and Bob likewise chooses a number b, both satisfying that $2 \le a, b \le p - 2$. For security reasons, it is essential that Alice keeps her number a to herself and that Bob also keeps his number b to himself.

(3) Now Alice computes $g^a \pmod{p}$ and sends this number c to Bob. Bob then computes $c^b \pmod{p}$. Notice that the net effect is that Bob (without knowing a) has then the value $g^{ab} \pmod{p}$, since

$$c^b \equiv (g^a)^b \equiv g^{ab} \pmod{p}.$$

(4) In the same way, Bob calculates $g^b \pmod{p}$ and sends this number d to Alice. She then computes $d^a \pmod{p}$, but again notice that she (without knowing b) has the same value

$$d^a \equiv (g^b)^a \equiv g^{ba} \pmod{p}.$$

(5) But now $g^{ab} \pmod{p} = g^{ba} \pmod{p}$, and we have arrived at a *secret common key*, namely the number $g^{ab} \pmod{p}$, which Alice and Bob can now use for secure communication. Note that an evil eavesdropper Eve cannot obtain this key from the information

exchange since she will only have seen the numbers c and d. Even if Eve knows p and g (which Alice and Bob had to agree on to get started), she cannot compute the common key since she does not know a and b. After a couple of examples we shall discuss this point further.

Example 6.3. We first illustrate the Diffie-Hellman key-exchange method using numbers which of course cannot be used in practice since they are *much* too small to be secure.

(1) Alice and Bob agree that their modulus p will be 11 and that their primitive element g will be 2. (We saw in Example 6.2 that 2 is a primitive root in $\mathbb{Z}_{11}$.)

(2) Alice chooses her secret number a to be 8 and Bob chooses his secret number b to be 9.

(3) Alice computes $2^8 \pmod{11} = 3$ and sends it to Bob. Upon receiving the value 3 from Alice, Bob calculates $3^9 \pmod{11} = 4$. Bob now knows the secret common key is 4.

(4) Meanwhile Bob computes $2^9 \pmod{11} = 6$ and sends it to Alice. Upon receipt, Alice computes $6^8 \pmod{11} = 4$. She now also knows that the common key is 4.

(5) Note that the eavesdropper Eve has only seen the 3 and the 6, from which she has no way of finding the 4 (thanks to the secrecy of the 8 and 9, even though she may have discovered the modulus 11 and the primitive root 2).

Let's look now at another example which is somewhat larger but still not close to large enough to be secure, showing how Alice and Bob might use a mathematical software program like *Mathematica*© to do the necessary calculations.

Example 6.4. (1) Alice and Bob decide to use a modulus p which lies between a million and ten million. According to the Prime Number Theorem (see Chapter 11), about one out of every $\ln(5,000,000) \approx 15$ numbers in this range is prime ($\ln$ is the natural log function), so it won't take too long to happen on one. They discover that $p = 2,134,879$ will work; for example, in *Mathematica*, they see

$$\text{PrimeQ}[2134879] = \text{True}.$$

Their software also can quickly identify a primitive root g, specifically, they see

$$\text{PrimitiveRoot}[2134879] = 6.$$

(2) Alice picks $a = 8,254$ for her secret number and Bob picks $b = 47,235$ for his.

(3) Alice's software will use fast exponentiation (as covered in Chapter 5) to compute any base (say g) raised to any power (say a) modulo any modulus (say p). So she computes

$$\text{PowerMod}[6,8254,2134879] = 808247,$$

which is the value c that she sends over to Bob. Bob now computes

$$\text{PowerMod}[808247,47235,2134879]= 1205481.$$

Hence he now knows that the shared secure key is 1,205,481.

(4) Likewise, Bob computes

$$\text{PowerMod}[6,47235,2134879] = 1981047$$

and sends this value d to Alice, who then computes

$$\text{PowerMod}[1981047,8254,2134879] = 1205481.$$

Now she also knows the shared key to be 1,205,481, but the eavesdropper Eve does not, and we are done.

You may be wondering how secure the Diffie-Hellman scheme really is. The numbers $c = g^a \in \mathbb{Z}_p$ and $d = g^b \in \mathbb{Z}_p$ are sent across the communication line, so Eve definitely has access to the values c and d. As we have noted, she may also know the prime modulus p and the primitive element g (since Alice and Bob had to somehow agree on them). Hence the real problem here, known as the *Discrete Logarithm Problem*, is the following:

Discrete Logarithm Problem. In the equation $g^a \pmod{p} = c$, if we know the base g, the modulus p where p is prime, and the answer c, can we discover the exponent $a \in \mathbb{Z}_p$?

This problem is considered by theoretical computer scientists to be very difficult to solve for large primes p. (See Problem 6.4 for a formula for the Discrete Logarithm Problem.) It is this difficulty which keeps an eavesdropper from being able to work out the secret key even if she intercepts p, g, and c.

6.5 Public Key Cryptography and the RSA System

Whereas the Diffie-Hellman key exchange method depends upon Fermat's Theorem (Theorem 5.1) and on the existence of primitive roots in the set $\mathbb{Z}_p$ where p is prime, the cryptographic system we shall now introduce, called the RSA system, depends instead on Euler's Function and Euler's Theorem (Theorem 5.4), as introduced in the previous chapter. The system is named after its founders Ronald Rivest, Adi Shamir, and Leonard Adleman; see [11]. Though introduced in the late 1970's, this system remains in wide use today for digital communications of all sorts, including in particular financial transactions such as on-line payments with a credit card.

An important new idea in the RSA system is that it involves *public keys.* This conceptual breakthrough showed that it is possible to avoid the dependence on private keys which themselves require secure exchange. As designed in RSA, the public keys are made possible by the fact that *factorization of integers is hard*, especially when the primes involved in the factorization are large.

So let us suppose that Alice now would like *anyone* to be able to send her a secure encrypted message which she and only she can decrypt. Here is her procedure:

(1) She selects two large primes p and q of the same approximate size and carefully keeps these choices private! In practice these primes need to have at least 100 decimal digits to guarantee security. She then computes her modulus $n = pq$.

(2) Now Euler's Function and Theorem come in. She computes

$\phi(n) = \phi(pq) = (p-1)(q-1)$ and keeps this value private! Now she selects a number e, called her *encrypting exponent*, which by using the Euclidean Algorithm she carefully checks is relatively prime to $\phi(n)$. (If it is not relatively prime to $\phi(n)$, she picks another e and checks it, etc.) As we saw in Lemma 3.3, this calculation can be run backwards to discover the multiplicative inverse d (called the *decrypting exponent*) of e modulo $\phi(n)$. That is, for some positive integer k, we have $ed - k\phi(n) = 1$, so that $ed = 1 + k\phi(n)$. Again, *she keeps the value of d very secret.*

(3) She publishes, for all the world (including the eavesdropper Eve) to see, her modulus n and her encrypting exponent e. This is why RSA is an example of *public key cryptography*. She keeps the values of p and q a secret, so in addition the values of $\phi(n)$ and in particular the decrypting exponent d are unknown to the outside world. She is now ready to receive messages encrypted via her public n and e which she alone can decrypt.

Now Bob would like to send a message m to Alice which will be encrypted to ensure its security. The message m needs to be *digitized* in some standard agreed-upon way (that is, if the message contains symbols other than just digits, it must be converted to all digits), and m must be a number which is positive but less than Alice's public modulus n. If m is larger than n, Bob breaks m into smaller "packets" and transmits each separately. (We'll assume here that m is small enough to send in full.) In addition, we shall assume that m and n are relatively prime in order to apply Euler's Theorem. (When $n = pq$ is large, the chances that n and m are *not* relatively prime are infinitesimal.) Bob now sends the following encrypted digitized message to Alice:

$$c = m^e \pmod{n}.$$

Upon receiving this message, Alice uses Euler's Theorem, which says in this case that $m^{\phi(n)} \equiv 1 \pmod{n}$. She applies her decrypting exponent d to c and reduces modulo n. Here is what happens, because of Euler's Theorem and the fact that $ed = 1 + k\phi(n)$ (for some k, as given in Step (2) above):

$$c^d \equiv m^{ed} = m^{1+k\phi(n)} = (m)(m^{\phi(n)})^k \equiv (m)(1^k) = m \pmod{n}.$$

We see then that Alice, using her secret d, has successfully decrypted Bob's message, which he encrypted using *her* public values n and e. Isn't RSA a truly elegant system? Again, it depends on two basic things: the difficulty of factoring large numbers, and Euler's Theorem. Eve the Eavesdropper would love to know the value of d, but she cannot compute it unless she can figure out p and q, which involves factoring the huge number n.

As we did with the Diffie-Hellman key exchange method, let's look at a couple of examples, the first using small numbers to illustrate the various calculations and the second with larger numbers (but not as large as needed in "real life" to guarantee security).

Example 6.5. First, Alice selects $p = 11$ and $q = 17$ for her two (secret) primes, so that her modulus is $n = (11)(17) = 187$. Next, she computes $\phi(n) = (11 - 1)(17 - 1) = 160$ and selects her encrypting exponent $e = 13$. Using the Euclidean Algorithm she verifies that 13 and 160 are relatively prime, and in fact she computes that $(13)(37) - 3(160) = 1$, which says that 37 is the multiplicative inverse of 13 modulo 160. Thus Alice's very secret decrypting exponent is $d = 37$. She now publishes her public modulus $n = 187$ and her public encrypting exponent $e = 13$.

Now Bob wants to secretly meet Alice at 8 tonight, so he wishes to send her the encrypted message $m = 8$. Using fast exponentiation (see Example 5.11), he computes and sends to her

$$c = m^e \pmod{n} = 8^{13} \pmod{187} = 94.$$

Upon receiving Bob's encrypted message $c = 94$, she uses fast exponentiation to decrypt it:

$$c^d \pmod{n} = 94^{37} \pmod{187} = 8,$$

so now she knows what time to meet Bob, but no one else does.

As mentioned above, all messages must be digitized, and in some standardized way. In practice what's often used is the so-called ASCII codes, which are a standardized conversion of letters and symbols into base-10 numbers between 0 and 127, but here let us just use a slightly simpler system as follows: "A" will be denoted by 01, "B" will be denoted by 02, ..., "Z" will be denoted by 26. We will likely want to have spaces, commas, etc., so a space will be denoted by 00, a comma by 27, a period by 28, and an exclamation point by 29.

Let's now do an example with more realistic numbers for security, but still not large enough to be used in practice. We show the functions which would be employed if we were doing the calculations in *Mathematica*.

Example 6.6. This time Alice selects as her primes ones with about 25 digits, making use of a *Mathematica* function which identifies prime numbers in the given range using advanced algorithms (not factorization!):

$$p=\text{RandomPrime}[\{10^{24}, 10^{25}\}]=5401712780740566847737779,$$

$$q=\text{RandomPrime}[\{10^{24}, 10^{25}\}]=8599036489738585051862237,$$

so her public modulus is

$$n = pq =$$
$$46449525308675415080549048730750651874434608351623.$$

She computes $\phi(n) = (p - 1)(q - 1)$:

$$\phi(n) = 46449525308675415080549034730001381395282708751608.$$

Now she randomly selects her encrypting exponent e to be 137 and uses the Euclidean Algorithm to verify that it is relatively prime to $\phi(n)$ and to identify the multiplicative inverse d of e modulo $\phi(n)$ (the *Mathematica* function used here is ExtendedGCD):

$$d = 47466668198646409571363977096351776608318809653449.$$

Finally, she publishes n and e.

Now Bob wants to send Alice the encrypted message "MATH IS FUN!" (Bob is a math nerd). He uses the scheme above to digitize his message:

130120080009190006211429. Since he knows that n has about 50 digits whereas his message has 24, he can send it all in one packet. Using Alice's public n and e, Bob uses fast exponentiation to send over

$$\text{PowerMod}[130120080009190006211429,e,n]$$

$$=14921831572988403080434060477279391151249521396635.$$

Upon receiving the message, Alice uses her decrypting exponent d to get

$$\text{PowerMod}[14921831572988403080434060477279391151249521396635,d,n]$$

$$=130120080009190006211429,$$

which she can easily translate to "MATH IS FUN!" (She nods in agreement.)

6.6 Security versus Authenticity

We finish this chapter by discussing the need for *authenticity* as well as security in modern digital communication. In Example 6.5 Alice decrypts Bob's message to meet him at 8, *but* how can she be sure that Bob sent the message? Perhaps it was the evil Eve who actually sent it and plans to trick her into giving up her decrypting exponent d when they meet. Alice would like to know that the message from Bob is authentic. Well, it turns out that RSA can also be used to establish authenticity as well as guarantee security, via what's called a *digital signature*. This can be done by having as the last packet in a message a "signature" which, unlike the main part of the message, is encoded using *the sender's public modulus and private decoding exponent*.

Here's how it works. Let us now denote Alice's public keys by n_A and e_A and her private decrypting key by d_A. Of course Bob can also have public and private keys which we shall denote by n_B, e_B, and d_B. Now Bob wants to send the message m and his signature

s to Alice. As before he uses *her* public modulus and encrypting exponent on the m, but for the signature part he uses *his* own public modulus and his private decrypting exponent. Hence Alice receives two numbers c and t, say, which are

$$c = m^{e_A} \pmod{n_A}; \; t = s^{d_B} \pmod{n_B}.$$

Now upon receipt of the pair (c, t), she can decrypt both as follows:

$$m = c^{d_A} \pmod{n_A}; \; s = t^{e_B} \pmod{n_B}.$$

If the message really was from Bob, the resulting digitized signature s, when "undigitized," should make sense. On the other hand, if the signature was actually from anyone besides Bob, what would come out of her computation for s, when undigitized, would definitely not be s, but rather something unrecognizable.

Example 6.7. In Example 6.5 Alice assumed that the message "8" about when to meet had come from Bob but couldn't be sure. Let us assume now that a general agreement between them is that when sending a message to the other, they will include a signature at the end. Remembering that these numbers are far too small to be secure, let's assume that Alice's keys are as in that exercise but are now labeled $p_A = 11$, $q_A = 17$, $n_A = 187$, $e_A = 13$, and $d_A = 37$. Without going into the details, let's assume that Bob's private primes are $p_B = 17$ and $q_B = 23$, so his public modulus is $n_B = 391$. He selects for his public encrypting exponent $e_B = 15$ and then computes that his private decrypting exponent is $d_B = 47$.

Now all the pieces are in place, and Bob can send his message $m = $ "8" to Alice, following that with his signature s, say "2" since "B" is second in the alphabet. So he sends over

$$c = m^{e_A} \pmod{n_A} = 8^{13} \pmod{187} = 94, \text{ and}$$

$$t = s^{d_B} \pmod{n_B} = 2^{47} \pmod{391} = 77.$$

Alice, upon receipt of the pair $(c = 94, t = 77)$, decrypts as follows:

$$c^{d_A} \pmod{n_A} = 94^{37} \pmod{187} = 8, \text{ and}$$

$$t^{e_B} \pmod{n_B} = 77^{15} \pmod{391} = 2.$$

Thus Alice knows that the message is both secure and authentic.

However, let's now suppose that Eve the Eavesdropper, whom Alice does not know, actually is the one who sends the message. Eve knows that Alice will be looking for Bob's signature, but she can't encrypt that because *she does not know his private d_B*, so she randomly picks a value, say 101, and sends the pair $(c = 94, t = 101)$ to Alice. But when Alice decrypts the signature, hoping the message is from Bob, she gets

$$t^{e_B} \pmod{n_B} = 101^{15} \pmod{391} = 305,$$

and she now knows that this message is definitely not from Bob. In fact Eve had a one out of 391 chance of picking the right number to send to Alice in order for the decryption of the signature to yield a "2", so Eve is foiled by RSA again!

Once the idea of public key encryption emerged, numerous other public key systems have been developed, some of which (for example "elliptic curve encryption") are more efficient than RSA in that they are secure using smaller keys. Nonetheless, RSA, the simplest and most elegant of these systems, remains in significant use. When you use your computer today to do any secure communication, RSA may well be working for you in the background.

6.7 Summary

Number theory plays a key role in modern computer science in general and in digital cryptography in particular. In this chapter we have studied two very important encryption ideas, the Diffie-Hellman key exchange method and the RSA public key encryption system. Both of these, when employed properly, ensure digital security, and both make fundamental use of number theory concepts including the Euclidean Algorithm, multiplicative inverses and primitive roots in $\mathbb{Z}_p$, fast exponentiation, Fermat's Theorem, Euler's Function, and Euler's Theorem. Because of number theory, we can (and do) all partake in secure digital communication every single day.

6.8   Solved Problems

Linear Encryption

6.1. Using the exact same linear encryption scheme as in Example 6.1, my broker sends me an encrypted reply to my "SELL" message. It translates as "EKMC." By decrypting, what is her message?

Solution:
The encoded message EKMC, translated into numbers, is the set $\{4, 10, 12, 2\}$. We now subtract 12 from each of these and then multiply by the inverse of 5 in $\mathbb{Z}_{26}$, which is 21. Modulo 26 this gives us

$$21(\{4 - 12, 10 - 12, 12 - 12, 2 - 12\}) \equiv -5(\{-8, -2, 0, -10\})$$

$$\equiv \{14, 10, 0, 24\},$$

which translates to OKAY.

Primitive Roots and the Diffie-Hellman Key Exchange Method

6.2. (a) Find the smallest primitive root in $\mathbb{Z}_{13}$.
(b) Assume that Alice and Bob are using the Diffie-Hellman key exchange method to create a common secure key and have agreed on 13 for the modulus and the answer of Part (a) as the primitive root. If Alice chooses her secret number to be $a = 3$ and Bob chooses his secret number to be $b = 5$, determine the common key.

Solution:
(a) As we saw in Example 6.3, since $\phi(13 - 1) = 4$, there will be four primitive roots in $\mathbb{Z}_{13}$. Moreover, if k is the smallest exponent on an element a of $\mathbb{Z}_{13}$ for which $a^k \equiv 1 \pmod{13}$, then k divides $13 - 1 = 12$. Hence, testing 2, we compute that $2^6 = 64 \equiv 12 \equiv -1 \pmod{13}$, so $2^{12} \equiv 1 \pmod{13}$, and 12 is the smallest such exponent, so 2 is a primitive root modulo 13.

(b) Alice sends $2^3 = 8$ to Bob and Bob, using Fermat's Theorem, computes $8^5 = 2^{15} \equiv 2^3 = 8 \pmod{13}$. On the other hand, Bob sends Alice $2^5 \equiv 6 \pmod{13}$ and Alice computes $6^3 \equiv 8 \pmod{13}$. Hence their common secure key is 8.

6.3. If you have access to some good mathematical software, re-peat Problem 6.2 with $p = 577$, $a = 52$ and $b = 34$. (Note: Here is a way to find the smallest primitive root modulo a prime p: Compute $x^{(p-1)/2} \pmod{p}$ for $x = 2, 3, \ldots$ until you find one which is *not* 1.)

Solution:
The smallest primitive root in $\mathbb{Z}_{577}$ is 5. Alice sends Bob the num-ber $5^{52} \equiv 428 \pmod{577}$. Bob sends Alice the number $5^{34} \equiv 53 \pmod{577}$. Their common secret key is $5^{(52)(34)} \equiv 130 \pmod{577}$.

6.4. In Section 1.4 we indicated that the Discrete Logarithm Prob-lem is viewed as a difficult problem. In spite of this being the case, there is however a formula to compute this number (see [9] for a proof of this formula).

If p is a prime and g is a primitive element in $\mathbb{Z}_p$, then for a non-zero in $\mathbb{Z}_p$

$$\log_g(a) = -1 + \sum_{j=1}^{p-2} \frac{a^j}{g^{-j} - 1}.$$

Why is this formula likely not useful when p is large?

Solution:
When p is a large prime (say with at least 100 digits), one cannot compute the formula since one would have to compute $p - 2$ sep-arate terms. So even though we have a valid formula, we cannot compute it!

Public Keys and RSA Encryption

6.5. Suppose you wish to use the RSA system to receive secure messages. You pick as your private primes $p = 17$ and $q = 23$.
(a) Compute your public modulus n and your private $\phi(n)$.
(b) Verify that $e = 5$ is a possible public encrypting exponent by using the Euclidean Algorithm applied to $\phi(n)$ and e, and find the decrypting exponent d by running the algorithm backwards. (Note: It may be helpful to review Theorem 1.3 and Examples 1.5 and 3.7 before proceeding.)

Solution:
(a) $n = 17 * 23 = 391$, $\phi(391) = \phi(17)\phi(23) = 16 * 22 = 352$.
(b) Running the Euclidean Algorithm forward:

$$352 = 5(70) + 2$$
$$5 = 2(2) + 1.$$

and now backwards:

$$1 = 5 - 2(2) = 5 - 2(352 - 5(70)) = 5(141) - 2(352).$$

Hence $5(141) \equiv 1 \pmod{352}$, so $d = 141$.

6.6. In the RSA system in Problem 6.5, how many different possible encrypting exponents e are there with $1 \le e < \phi(n)$?

Solution:
The encrypting exponent e must be relatively prime to $\phi(17*23) = 16 * 22 = 352$, and we know that the number of elements of $\mathbb{Z}_{352}$ which are relatively prime to 352 is $\phi(352) = \phi(32)\phi(11) = 16 * 10 = 160$.

6.7. Your friend Chuck wants to send you the secret message 4. He uses your public modulus n and encrypting exponent e from Problem 6.5. What encrypted message will you receive from Chuck?

Solution:
You receive 242, since $4^5 = 1024 \equiv 242 \pmod{391}$.

Security versus Authenticity

6.8. Referring to Problems 6.5 and 6.7, suppose you want to send a message back to Chuck to which you wish to append with your digital signature, a number between 1 and 10. Chuck receives the number 9 and decrypts it. What was your signature?

Solution:
When Chuck decrypts the message 9, which you encoded using your public modulus 391 and your *private decrypting exponent* $d = 141$, he will use your modulus and your public encrypting exponent 5. Hence he computes $9^5 = 59,049 \equiv 8 \pmod{391}$, i.e., your signature was 8.

6.9   Supplemental Problems

Linear Encryption

6.9. Using the exact same linear encryption scheme as in Example 6.1, my broker sends me an encrypted reply to my "SELL" message. It translates as "BEZG." By decrypting, what is her message?

Primitive Roots and the Diffie-Hellman Key Exchange Method

6.10. (a) Find the smallest primitive root in $\mathbb{Z}_{17}$. (See Example 6.3 and Problem 6.2.)
(b) Assume that Alice and Bob are using the Diffie-Hellman key exchange method to create a common secure key and have agreed on 17 for the modulus and the answer of Part (a) as the primitive root. If Alice chooses her secret number to be $a = 4$ and Bob chooses his secret number to be $b = 7$, determine the common key.

6.11. If you have access to some good mathematical software, repeat Problem 6.10 with $p = 431$, $a = 25$ and $b = 92$. (The note on Problem 6.3 may be helpful.)

6.12. Question: If g is a primitive root modulo p, is $p - g$ also a primitive root modulo p? Let's examine some data, assuming $p > 2$.

(a) We have seen that 2 is a primitive root modulo 11. Is it true or false that $11 - 2 = 9$ is also a primitive root modulo 11?

(b) We have also seen that 2 is a primitive root modulo 13. Is it true or false that $13 - 2 = 11$ is also a primitive root modulo 13?

(c) See if you can formulate a conjecture about what must be true of $p - 1$ in order for g being a primitive root modulo p to imply that $p - g$ is also a primitive root modulo p. (Hint: If g is primitive, think of $p - g$ as $-g$. When will $g^{(p-1)/2}$ be the same as $(-g)^{(p-1)/2}$?)

Public Keys and RSA Encryption

6.13. Suppose you wish to use the RSA system to receive secure messages. You pick as your private primes $p = 19$ and $q = 31$.

(a) Compute your public modulus n and your private $\phi(n)$.

(b) Verify that $e = 7$ is a possible public encrypting exponent by using the Euclidean Algorithm applied to $\phi(n)$ and e, and find the decrypting exponent d by running the algorithm backwards. (Note: It may be helpful to review Theorem 1.3 and Problem 6.5 before proceeding.)

6.14. (a) In the RSA system in Problem 6.13, how many different possible encrypting exponents e are there with $1 \leq e < \phi(n)$?

(b) More generally, if $n = pq$ is *any* RSA modulus, how many possible encrypting exponents e are there as a function of n?

6.15. Your friend Sally wants to send you the secret message 6. She uses your public modulus n and encrypting exponent e from Problem 6.13. What encrypted message will you receive from Sally?

6.16. If you and a friend have access to good mathematical software, set up an RSA-enabled secure communication channel as follows:

(1) You each privately pick two primes p and q with $5200 < p, q < 10000$. Compute your modulus n, your encrypting exponent e, and

your decrypting exponent d. Share your values of n and e with each other.

(2) Using the digitizing scheme described prior to the Example 6.6, send to each other a digitized encrypted message followed by an encrypted digital signature. Each of these must consist of between 1 and 4 characters (so the digitized message and signature each has at most 8 digits).

(3) Decrypt the incoming messages and signatures, and compare notes. Good luck!

6.17. Concerning the difficulty of factoring larger numbers, especially "by hand," it can help to have some special information about the factors. For example, without the aid of mathematical software see if you can factor $n = 246,973$ given the following information: n is the product of two primes p and q which differ by 12. (Hint: What is $\sqrt{n}$?)

Security versus Authenticity

6.18. Referring to Problems 6.13 and 6.15, suppose you want to send a message back to Sally to which you wish to append with your digital signature, a number between 1 and 10. Sally receives the number 489 and decrypts it. What was your signature?

Answers to Selected Supplementary Problems

6.9. DONE.

6.10. (a) 3, (b) 4.

6.11. (a) 7, (b) 144.

6.12. (a) No, (b) Yes.

6.13. (a) $n = 19 * 31 = 589$, $\phi(n) = 18 * 30 = 540$, (b) $d = 463$.

6.14. (a) $\phi(\phi(19 * 31)) = \phi(18 * 30) = \phi(4)\phi(27)\phi(5) = 2 * 18 * 4 = 144$.

6.15. 162.

6.18. 3.

Chapter 7

Quadratic Residues and Quadratic Reciprocity

7.1 Introduction

If p is an odd prime number and a is a non-zero integer, does the quadratic congruence $x^2 \equiv a \pmod{p}$ have any solutions in $\mathbb{Z}_p$? The answer is that it depends on a. If there are solutions, we shall call a a "quadratic residue modulo p." The theory of what elements are or are not quadratic residues turns out to be very interesting, so we shall explore this theory now. Along the way we shall encounter two fundamental results in number theory: Gauss's Lemma, which gives an effective method for identifying quadratic residues, and the Law of Quadratic Reciprocity, which describes the relationship of the "quadratic character" of two odd primes p and q with respect to each other. This is not easy material, but the central results are very nice and, we believe, well worth the effort.

7.2 Quadratic Residues and the Legendre Symbol

Definition 7.1. Assume that the positive integers a and n are relatively prime. The integer a is a *quadratic residue modulo n* if

DOI: 10.1201/9781003193111-7

the congruence $x^2 \equiv a \pmod{n}$ has a solution. If the congruence does not have a solution, a is a *quadratic non-residue modulo n*.

We note that since the concepts of quadratic residue and non-residue are defined in terms of congruence modulo n, it is sufficient to consider only those residues or non-residues which are distinct modulo n; that is, we can restrict ourselves to non-zero elements of $\mathbb{Z}_n$, the set of integers modulo n introduced in Chapter 3.

Example 7.1. In $\mathbb{Z}_5$, we have $1^2 = 1$, $2^2 = 4$, $3^2 = 4$, and $4^2 = 1$, and so 1 and 4 are quadratic residues modulo 5, while 2 and 3 are quadratic non-residues modulo 5. Similarly in $\mathbb{Z}_7$, $1^2 = 6^2 = 1$, $2^2 = 5^2 = 4$, and $3^2 = 4^2 = 2$, so the integers 1, 2, and 4 are quadratic residues, while 3, 5, and 6 are quadratic non-residues modulo 7.

As this example illustrates, if p is an odd prime and if $j \in \mathbb{Z}_p$ is a solution of $x^2 \equiv a \pmod{p}$, then $p - j$ is also a solution, and $p - j \neq j$, for if they were equal, we would have that $p = 2j$, i.e., p would be even. Hence in $\mathbb{Z}_p$ the congruence $x^2 \equiv a \pmod{p}$ has either two distinct solutions or no solutions.

We now define the Legendre symbol, named in honor of Adrian-Marie Legendre (1752 - 1833).

Definition 7.2. Let p be an odd prime and let a be an integer which is relatively prime to p. The *Legendre symbol* $\left(\frac{a}{p}\right)$ is defined to be 1 if a is a quadratic residue modulo p, and -1 if a is a quadratic non-residue modulo p.

Example 7.2. Following up on Example 7.1, we have $\left(\frac{4}{5}\right) = 1$ but $\left(\frac{2}{5}\right) = -1$. Similarly, $\left(\frac{2}{7}\right) = 1$ but $\left(\frac{3}{7}\right) = -1$.

Theorem 7.2 below summarizes a few basic properties of the Legendre symbol. However, for the theorem's proof we first need the following result, named in honor of John Wilson (1741 - 1793).

Lemma 7.1. (*Wilson's Theorem*) *If p is an odd prime, then $(p - 1)! \equiv -1 \pmod{p}$.*

Proof. In $\mathbb{Z}_p$, if $a^2 = 1$, i.e., if a is its own multiplicative inverse, then $(a-1)(a+1) = 0$. Since p is prime, both $a-1$ and $a+1$ are relatively prime to p and hence either can be divided out of $(a-1)(a+1) = 0$ by Lemma 3.2. Thus $a = 1$ or $a = p-1$. For all the other elements of $\mathbb{Z}_p$ (i.e., $\{2, 3, \ldots, p-2\}$), when we pair each element up with its multiplicative inverse and then multiply all these pairs together, we get $(p-2)! \equiv 1 \pmod{p}$. Now multiplying through by $p-1$, we get $(p-1)! \equiv p-1 \equiv -1 \pmod{p}$, as desired. $\square$

Theorem 7.2. *Let p be an odd prime and let a and b be integers which are relatively prime to the prime p. Then*
  (a) $\left(\frac{a}{p}\right) \equiv a^{(p-1)/2} \pmod{p}$,
  (b) $\left(\frac{a}{p}\right)\left(\frac{b}{p}\right) = \left(\frac{ab}{p}\right)$,
  (c) *If $a \equiv b \pmod{p}$, then* $\left(\frac{a}{p}\right) = \left(\frac{b}{p}\right)$,
  (d) $\left(\frac{a^2}{p}\right) = 1, \left(\frac{1}{p}\right) = 1, \left(\frac{-1}{p}\right) = (-1)^{(p-1)/2}$.

Proof. (a) If $\left(\frac{a}{p}\right) = 1$, then the congruence $x^2 \equiv a \pmod{p}$ has a solution x_0. Then, by Fermat's Theorem (Theorem 5.1),

$$a^{(p-1)/2} \equiv (x_0^2)^{(p-1)/2} \equiv x_0^{p-1} \equiv 1 \equiv \left(\frac{a}{p}\right) \pmod{p}.$$

On the other hand, if $\left(\frac{a}{p}\right) = -1$, then the congruence $x^2 \equiv a \pmod{p}$ has no solution. Thus, for each integer j with $1 \leq j \leq p-1$, we associate the unique integer i so that $ji \equiv a \pmod{p}$ with $0 \leq i \leq p-1$. Specifically, in $\mathbb{Z}_p$, $i = j^{-1}a$, and $i \neq 0$. In addition, $i \neq j$ since if it were, we would have $j^2 \equiv ji \equiv j(j^{-1}a) \equiv a \pmod{p}$, i.e., j would be a solution of $x^2 \equiv a \pmod{p}$. Hence the integers $1, \ldots, p-1$ can be paired off, j and its associate i, and there are $(p-1)/2$ such pairs. If we now multiply all of these pairs together, we obtain $(p-1)! \equiv a^{(p-1)/2} \pmod{p}$. Hence from Wilson's Theorem (Lemma 7.1) we obtain that $a^{(p-1)/2} \equiv -1 \equiv \left(\frac{a}{p}\right) \pmod{p}$. The proof of Part (a) is now complete.

   The remaining parts of the theorem are all simple consequences of Part (a). See Problem 7.17. $\square$

7.3   Computing the Legendre Symbol

Theorem 7.2 tells us important facts about the Legendre symbol. Here we state and prove two theorems which give us new insight into this symbol and new ways to compute it. We first need a lemma which follows from Fermat's Theorem (Theorem 5.1).

Lemma 7.3. *If p is an odd prime and $\gcd(a,p) = 1$, then the congruence $x^2 \equiv a \pmod{p}$ has two solutions or no solutions according as $a^{(p-1)/2} \equiv \pm 1 \pmod{p}$.*

Proof. Fermat's Theorem implies that

$$(a^{(p-1)/2} - 1)(a^{(p-1)/2} + 1) \equiv a^{p-1} - 1 \equiv 0 \pmod{p}.$$

Hence we have that $a^{(p-1)/2} \equiv \pm 1 \pmod{p}$. The result now follows from Theorem 7.2 Part (a) and from the remarks directly following Example 7.1. □

We now state and prove a theorem which gives us a specific way to compute the Legendre symbol for a given prime modulus p and a given positive integer a for which $\gcd(a,p) = 1$.

Theorem 7.4. (*Gauss's Lemma*) *Let p be an odd prime and let a be an integer relatively prime to p. Consider the set $\{a, 2a, 3a, \ldots, ((p-1)/2)a\} \pmod{p}$ of elements of $\mathbb{Z}_p$. If n denotes the number of these elements which are greater than $p/2$, then $\left(\frac{a}{p}\right) = (-1)^n$.*

Proof. Let $r_1, \ldots, r_n$ denote the set of elements that exceed $p/2$; let $s_1, \ldots, s_k$ denote the set of remaining elements. The r_i and the s_i are clearly distinct, and moreover, none of them is 0. In addition $n + k = (p-1)/2$. For $i = 1, \ldots, n, 0 < p - r_i < p/2$, and the values $p - r_i$ are distinct.

We now observe that no value $p - r_i$ can equal an s_j. Suppose $p - r_i = s_j$, then $r_i \equiv ua, s_j \equiv va$ for $1 \leq u \leq (p-1)/2, 1 \leq v \leq (p-1)/2$, and then $p - ua \equiv va \pmod{p}$. We have $a(u+v) \equiv 0 \pmod{p}$, and since $(a,p) = 1$, we get $u + v \equiv 0 \pmod{p}$, which is a contradiction.

Thus the values $p - r_1, \ldots, p - r_n, s_1, \ldots, s_k$ are all distinct. In addition, each value is at least 1 and each value is less than

$p/2$. And there are $n + k = (p-1)/2$ such values. These numbers are thus a rearrangement of the numbers $1, \ldots, (p-1)/2$ in some order. Multiplying them together we obtain

$$(p - r_1) \cdots (p - r_n) s_1 \cdots s_k = 1 \cdot 2 \cdots \frac{p-1}{2}.$$

Hence

$$(-r_1) \cdots (-r_n) s_1 \cdots s_k \equiv 1 \cdot 2 \cdots \frac{p-1}{2} \pmod{p},$$

$$(-1)^n r_1 \cdots r_n s_1 \cdots s_k \equiv 1 \cdot 2 \cdots \frac{p-1}{2} \pmod{p},$$

$$(-1)^n a \cdot 2a \cdots \frac{p-1}{2} a \equiv 1 \cdot 2 \cdots \frac{p-1}{2} \pmod{p}.$$

The factors $2, 3, \cdots, (p-1)/2$ can be cancelled to obtain $(-1)^n a^{(p-1)/2} \equiv 1 \pmod{p}$. Hence we have $(-1)^n \equiv a^{(p-1)/2} \equiv \left(\frac{a}{p}\right)$ $\pmod{p}$, and the proof is complete. $\square$

Example 7.3. Suppose $p = 11$ and $a = 7$. The question is, Is 7 a quadratic residue modulo 11 or not? Since $(11-1)/2 = 5$, we compute

$$\{7, 2 \cdot 7, 3 \cdot 7, 4 \cdot 7, 5 \cdot 7\} = \{7, 14, 21, 28, 35\} \equiv \{7, 3, 10, 6, 2\} \pmod{11}.$$

Since 6, 7 and 10 exceed $11/2 = 5.5$, by Gauss's Lemma we obtain that $\left(\frac{7}{11}\right) = (-1)^3 = -1$, and so 7 is a quadratic non-residue modulo 11. In fact, one can compute that the set of quadratic residues modulo 11 is $\{1, 4, 9, 5, 3\}$.

We now state and prove another theorem which concerns a method for computing the Legendre symbol. For this we need the concept of "the greatest integer of a real number." If r is a real number, the greatest integer less than or equal to r will be denoted by $[r]$. Thus, for example, $[7/2] = 3$, $[15] = 15$, and $[-4/3] = -2$. The greatest integer function will shortly play an important role in the theory of quadratic reciprocity. Be sure not to confuse this notation with the notation for the Legendre symbol. For example, the greatest integer $[7/11]$ is 0; the Legendre symbol $\left(\frac{7}{11}\right)$ is -1 (as seen in Example 7.3).

Theorem 7.5. *Let p be an odd prime and assume that* $\gcd(a, 2p) = 1$. *Then* $\left(\frac{a}{p}\right) = (-1)^t$ *where* $t = \sum_{j=1}^{(p-1)/2}[ja/p]$ *and* $\left(\frac{2}{p}\right) = (-1)^{(p^2-1)/8}$.

Proof. We will employ the same notation as in the proof of Gauss' Lemma (Theorem 7.4) above. The r_i and s_i are the least positive remainders obtained by dividing the integers ja by the prime p for $j = 1, 2, \ldots, (p-1)/2$; the quotients are $[ja/p]$ for each j. We have

$$\sum_{j=1}^{(p-1)/2} ja = \sum_{j=1}^{(p-1)/2} p[ja/p] + \sum_{j=1}^{n} r_j + \sum_{j=1}^{k} s_j;$$

that is, for each j we have $ja = $ divisor times quotient $+$ remainder. In addition, because $\gcd(a, p) = 1$ and by the proof of Theorem 7.4, we have

$$\sum_{j=1}^{(p-1)/2} j = \sum_{j=1}^{n}(p - r_j) + \sum_{j=1}^{k} s_j = np - \sum_{j=1}^{n} r_j + \sum_{j=1}^{k} s_j.$$

Upon subtracting the left- and right-hand expressions of the second line from the left- and right-hand expressions of the first line we obtain

$$(a - 1)\sum_{j=1}^{(p-1)/2} j = p\left(\sum_{j=1}^{(p-1)/2} [ja/p] - n\right) + 2\sum_{j=1}^{n} r_j.$$

We note (using the standard formula $\sum_{i=1}^{k} i = k(k+1)/2$) that

$$\sum_{j=1}^{(p-1)/2} j = \frac{p^2 - 1}{8},$$

and thus we obtain

$$(a - 1)\frac{p^2 - 1}{8} \equiv \sum_{j=1}^{(p-1)/2} [ja/p] - n \pmod{2}.$$

If a is odd, the left-hand side of the congruence is 0, which shows that $n \equiv \sum_{j=1}^{(p-1)/2}[ja/p]$ (mod 2). If $a = 2$ we see that $n \equiv (p^2 - 1)/8$ (mod 2) since we have $[2j/p] = 0$ for $1 \leq j \leq (p-1)/2$. Our result now follows from Gauss's Lemma (Theorem 7.4), and we are done. $\square$

Example 7.4. (a) In Example 7.3 we used Gauss's Lemma (Theorem 7.4) to see that 7 is a quadratic non-residue modulo 11 since the value of n there is odd (specifically $n = 3$). Theorem 7.5 now tells us that the sum of greatest integers $[ja/p]$ must also be odd. In fact, that sum is

$$[7/9] + [14/9] + [21/9] + [28/9] + [35/9] = 0 + 1 + 2 + 3 + 3 = 9,$$

and of course $(-1)^3 = (-1)^9 = -1$.

(b) We noted in Example 7.3 that 2 is also a quadratic non-residue modulo 11. By Theorem 7.5, we have $(p^2-1)/8 = (121-1)/8 = 15$, which is of course odd, so we again conclude that 2 in a non-residue.

7.4 Quadratic Reciprocity

Suppose p and q are both odd primes. If q is a quadratic residue modulo p, is p then a quadratic residue modulo q? The answer is no, but there is a clear relationship of their "quadratic characters" with respect to each other, and that relationship can be concisely stated in the following famous and elegant result.

Theorem 7.6. (*The Gaussian Reciprocity Law*) *If p and q are distinct odd primes,*

$$\left(\frac{q}{p}\right)\left(\frac{p}{q}\right) = (-1)^{\frac{p-1}{2}\frac{q-1}{2}}.$$

Before proving this result, we simply ask, isn't this a beautiful result? Let us discuss what, in essence, it tells us about the relationship of $\left(\frac{p}{q}\right)$ and $\left(\frac{q}{p}\right)$ when p and q are distinct odd prime numbers. We know that if an exponent k on -1 is even then $(-1)^k = 1$;

and if k is odd, then $(-1)^k$ is -1. Hence the right-hand side of our theorem's equation above will be -1 only if *both* $(p-1)/2$ and $(q-1)/2$ are odd; otherwise the right-hand side is 1. We can thus conclude the following:

(1) If both $(p-1)/2$ and $(q-1)/2$ are odd, then $\left(\frac{q}{p}\right) = -\left(\frac{p}{q}\right)$.

(2) Otherwise, $\left(\frac{q}{p}\right) = \left(\frac{p}{q}\right)$.

Example 7.5. (a) What is the relationship of $\left(\frac{17}{23}\right)$ and $\left(\frac{23}{17}\right)$? Well, $(17-1)/2 = 8$, so we already know that $\left(\frac{17}{23}\right) = \left(\frac{23}{17}\right)$.

(b) What is the relationship of $\left(\frac{19}{23}\right)$ and $\left(\frac{23}{19}\right)$? Well, $(19-1)/2 = 9$ and $(23-1)/2 = 11$, so we know that $\left(\frac{19}{23}\right) = -\left(\frac{23}{19}\right)$.

There have been many proofs of this result; in fact there are now at least 240 "different" proofs of this result; for example, google "proofs of Quadratic Reciprocity." We will not explain here what is meant by "different" proofs; rather, we provide the following proof of this very important result.

Proof. Let S be the set of all pairs of integers (x, y) which satisfy

$$1 \le x \le (p-1)/2, 1 \le y \le (q-1)/2.$$

The set S has $\frac{p-1}{2}\frac{q-1}{2}$ pairs of numbers. We note that there are no pairs (x, y) in S with $qx = py$ (since, for example, neither p nor y is divisible by q), so we now split the set S into two disjoint subsets S_1 and S_2 depending upon whether $qx > py$ or $qx < py$. We define the set S_1 by

$$S_1 = \{(x, y)| 1 \le x \le (p-1)/2, 1 \le y < qx/p\},$$

so the number of pairs in S_1 is given by $\sum_{x=1}^{(p-1)/2}[qx/p]$. Similarly we define the set S_2 by

$$S_2 = \{(x, y)| 1 \le y \le (q-1)/2, 1 \le x < py/q\},$$

and the number of pairs in S_2 is $\sum_{y=1}^{(q-1)/2}[py/q]$. Hence

$$\sum_{x=1}^{(p-1)/2} [qx/p] + \sum_{y=1}^{(q-1)/2} [py/q] = \frac{p-1}{2}\frac{q-1}{2}.$$

Finally, we apply Theorem 7.5 to p and q, obtaining

$$= (-1)^{\frac{p-1}{2}\frac{q-1}{2}} = (-1)^{\sum_{x=1}^{(p-1)/2}[qx/p]+\sum_{y=1}^{(q-1)/2}[py/q]} = (\frac{q}{p})(\frac{p}{q}),$$

and the proof is complete. □

This theorem was conjectured by Euler and Legendre but first proved by Gauss [1]. Gauss refers to it as the "fundamental theorem" in the Disquisitiones Arithmeticae and his papers, writing that "the fundamental theorem must certainly be regarded as one of the most elegant of its type." (Art. 151). Privately he referred to it as the "golden theorem" [2]. He published six proofs, and two more were found in his posthumous papers.

Example 7.6. Let us compute $(\frac{-42}{61})$. First, by Theorem 7.2 Part (b) we have

$$(\frac{-42}{61}) = (\frac{-1}{61})(\frac{2}{61})(\frac{3}{61})(\frac{7}{61}).$$

By Theorem 7.2 Part (d),

$$(\frac{-1}{61}) = (-1)^{60/2} = 1,$$

and by Theorem 7.5

$$(\frac{2}{61}) = (-1)^{(61^2-1)/8} = (-1)^{(3721-1)/8} = (-1)^{465} = -1.$$

Now by Quadratic Reciprocity, since $(61-1)/2 = 30$, which is even, and then reducing 61 modulo 3, we get

$$(\frac{3}{61}) = (\frac{61}{3}) = (\frac{1}{3}) = 1,$$

and similarly (since $(5-1)/2$ is also even, and again using Theorem 7.5)

$$(\frac{7}{61}) = (\frac{61}{7}) = (\frac{5}{7}) = (\frac{7}{5}) = (\frac{2}{5}) = (-1)^{24/8} = -1.$$

Hence $(\frac{-42}{61}) = 1 \cdot -1 \cdot 1 \cdot -1 = 1$.

To solve this problem more quickly, but still using Quadratic Reciprocity, one could also compute

$$(\frac{-42}{61}) = (\frac{19}{61}) = (\frac{61}{19}) = (\frac{4}{19}) = 1.$$

These techniques involving quadratic residues are also useful in solving the following kinds of problems.

Example 7.7. Let us describe all primes $p > 3$ so that 3 is a quadratic residue modulo p. By quadratic reciprocity and since $(3-1)/2 = 1$, we know that

$$(\frac{3}{p})(\frac{p}{3}) = (-1)^{(p-1)/2}.$$

Hence $(\frac{3}{p})$ will be 1 (i.e., 3 will be a quadratic residue modulo p) provided that $(\frac{p}{3}) = (-1)^{(p-1)/2}$; i.e., the two numbers $(\frac{p}{3})$ and $(-1)^{(p-1)/2}$ must either be both $+1$ or both -1. Now $(\frac{p}{3})$ equals $+1$ if $p \equiv 1 \pmod 3$ and equals -1 if $p \equiv 2 \pmod 3$. Moreover, $(-1)^{(p-1)/2}$ equals $+1$ if $p \equiv 1 \pmod 4$ and equals -1 if $p \equiv 3 \pmod 4$. We can conclude then that $(\frac{3}{p}) = 1$ provided that:

    (a) $p \equiv 1 \pmod 3$ and $p \equiv 1 \pmod 4$, or

    (b) $p \equiv 2 \pmod 3$ and $p \equiv 3 \pmod 4$,

which, by the Chinese Remainder Theorem (Theorem 4.2), is equivalent to $p \equiv 1$ or $11 \pmod{12}$.

7.5 Composite Moduli and the Jacobi Symbol

We shall finish this chapter by briefly considering the theory of quadratic residues and quadratic reciprocity for moduli which are not necessarily prime. We begin with

Definition 7.3. Suppose that m and n are relatively prime integers with n a positive odd number. Suppose that n's prime factorization is $n = p_1 \cdots p_s$, where the primes p_i are not necessarily distinct. Then the *Jacobi symbol* $(\frac{m}{n})$ is defined by

$$(\frac{m}{n}) = \prod_{j=1}^{s}(\frac{m}{p_j}),$$

where $(\frac{m}{p_j})$ is the Legendre symbol.

Two natural questions arise from this definition:
(i) Can the Jacobi symbol $(\frac{m}{n})$ be used to determine whether m is a quadratic residue or a quadratic non-residue modulo n?
(ii) Does the Jacobi symbol satisfy a "generalized quadratic reciprocity?"

The answer to (ii) is "yes" provided that we restrict m, like n, to be positive and odd. The answer to (i) is also "yes," but in a somewhat more complicated way than for the Legendre symbol. We now state these results without proof.

Theorem 7.7. *Suppose that m and n are relatively prime integers with n a positive odd number. Suppose that n's prime factorization is $n = p_1 \cdots p_s$, where the primes p_i are not necessarily distinct.*
(i)
    (a) *If the Jacobi symbol $(\frac{m}{n}) = -1$, then m is a quadratic non-residue modulo n.*
    (b) *If, for every $1 \leq j \leq s$, the Legendre symbol $(\frac{m}{p_j}) = 1$, then m is a quadratic residue modulo n.*
(ii)(*Generalized Quadratic Reciprocity*) *Suppose we assume further that m is also positive and odd. Then*

$$(\frac{m}{n})(\frac{n}{m}) = (-1)^{((m-1)/2)((n-1)/2)}.$$

Example 7.8. Let's examine these results for the cases of moduli $n = 9$ and $n = 15$. By direct computation in $\mathbb{Z}_n$ one can find that the non-zero quadratic residues in $\mathbb{Z}_9$ are $\{1, 4, 7\}$ and in $\mathbb{Z}_{15}$ are $\{1, 4, 9\}$. We consider the following three computations:

$$\left(\frac{7}{9}\right) = \left(\frac{7}{3}\right)\left(\frac{7}{3}\right) = \left(\frac{1}{3}\right)\left(\frac{1}{3}\right) = 1^2 = 1 \ (*)$$

$$\left(\frac{5}{9}\right) = \left(\frac{5}{3}\right)\left(\frac{5}{3}\right) = \left(\frac{2}{3}\right)\left(\frac{2}{3}\right) = (-1)^2 = 1 \ (**)$$

$$\left(\frac{11}{15}\right) = \left(\frac{11}{3}\right)\left(\frac{11}{5}\right) = \left(\frac{2}{3}\right)\left(\frac{1}{5}\right) = (-1)(1) = -1. \ (***)$$

Computation (*) illustrates the case of all Legendre symbols having value 1, which indicates that 7 is a quadratic residue modulo 9. Computation (**) illustrates the case of the Jacobi symbol having value 1 but the upper number 5 nonetheless being a non-residue modulo 9, due to the existence of an even number of -1's among the values of the Legendre symbols. Finally, Computation (***) illustrates the case of a value of -1 of the Jacobi symbol guaranteeing that 11 is a non-residue modulo 15.

Doing one example of generalized quadratic reciprocity, consider

$$\left(\frac{7}{15}\right) = \left(\frac{7}{3}\right)\left(\frac{7}{5}\right) = \left(\frac{1}{3}\right)\left(\frac{2}{5}\right) = (1)(-1) = -1 \text{ and } \left(\frac{15}{7}\right) = \left(\frac{1}{7}\right) = 1.$$

Hence the left-hand side $\left(\frac{7}{15}\right)\left(\frac{15}{7}\right)$ of the formula in Theorem 7.7, Part (ii) has value -1. Meanwhile the right-hand side also has value $(-1)^{((7-1)/2)((15-1)/2)} = (-1)^{21} = -1$, as guaranteed by the theorem.

7.6   Summary

At the beginning of this chapter we asked if the quadratic congruence $x^2 \equiv a \pmod{p}$ has solutions in $\mathbb{Z}_p$, where p is an odd prime. We now know the answer in some detail; specifically, any such congruence has either no solutions or two solutions, depending on whether $a \pmod{p}$ is a quadratic residue modulo p (of which there are $(p-1)/2$ in $\mathbb{Z}_p$) or a quadratic non-residue modulo p (of which there are also $(p-1)/2$ in $\mathbb{Z}_p$). The main question then becomes, How do we know which elements a of $\mathbb{Z}_p$ are of one type and which the other? This led us to define the Legendre symbol $\left(\frac{a}{p}\right)$, whose value is $+1$ if a is a quadratic residue and -1 otherwise. Using this symbol as a tool, we then developed various criteria for determining its value, and that is the content of Theorems 7.2, 7.4 (Gauss's Lemma), and 7.5.

We then turned to the question of quadratic reciprocity; that is, if p and q are both odd primes, how are $\left(\frac{q}{p}\right)$ and $\left(\frac{p}{q}\right)$ related? Theorem 7.6 (The Gaussian Reciprocity Law) gives a lovely, simple answer to that question. Finally, we discussed briefly the situation with non-prime moduli, and we learned that an analogous reciprocity theorem holds provided that we define the analogous Jacobi symbol $\left(\frac{m}{n}\right)$ in terms of the prime factorization of n and the corresponding Legendre symbols. As we said at the outset, this material is not easy, but the results we have been able to attain are both very important and quite elegant.

7.7 Solved Problems

Quadratic Residues and the Legendre Symbol

7.1. Prove that 3 is a quadratic residue modulo 13 but is a quadratic non-residue modulo 7.

Solution:
We have $4^2 \equiv 3 \pmod{13}$. The quadratic residues in $\mathbb{Z}_{13}$ are $1, 2, 4$ while the quadratic non-residues in $\mathbb{Z}_7$ are $3, 5, 6$.

7.2. Find all of the 8 quadratic residues in $\mathbb{Z}_{17}$.

Solution:
Obviously 1, 4, 9, and 16 are residues. After these we need to reduce modulo 17. For example then, $5^2 = 25 \equiv 8$. The remaining three residues are 2, 15, and 13.

7.3. Use the definition of the Legendre symbol (Definition 7.2) to find each of the following values:
(a) $(\frac{3}{5})$, (b) $(\frac{5}{11})$, (c) $(\frac{-1}{7})$, (d) $(\frac{4}{p})$ (p any odd prime).

Solution:
(a) -1 since 1 and 4 are the only quadratic residues modulo 5,
(b) 1 since $4^2 \equiv 5 \pmod{11}$,
(c) -1 since $-1 \equiv 6 \pmod 7$, but the quadratic residues modulo 7 are $\{1, 4, 2\}$,
(d) 1, since 4 is 2^2 for all $p \geq 5$, and $4 \equiv 1 \pmod 3$.

7.4. Using Theorem 7.2, find the values of
(a) $(\frac{100}{17})$, (b) $(\frac{2}{41})(\frac{18}{41})$, (c) $(\frac{42}{43})$.

Solution:
(a) By Theorem 7.2 Part (c), reducing 100 modulo 17, we obtain 15. Now see Problem 7.2, so the answer is 1.
(b) By Theorem 7.2 Part (b) we are evaluating $(\frac{36}{41})$, which is 1 since 36 is a perfect square.
(c) We are evaluating $(\frac{-1}{43})$, which by Theorem 7.2 Part (d) is $(-1)^{(43-1)/2} = (-1)^{21} = -1$.

Computing the Legendre Symbol

7.5. Determine the values of $[3/2], [-3/2], [37/11], [-39/10]$, where $[m]$ denotes the greatest integer in the real number m.

Solution:
    The values are $1, -2, 3$, and -4.

7.6. We know that 3 is a quadratic residue modulo 13 since $4^2 \equiv 3 \pmod{13}$. Compute the Legendre symbol $(\frac{3}{13})$ three different ways presented to us in this chapter:
(a) Use Theorem 7.2 Part (a) to compute $(\frac{3}{13})$,
(b) Use Theorem 7.4 (Gauss's Lemma) to compute $(\frac{3}{13})$,
(c) Use Theorem 7.5 to compute $(\frac{3}{13})$.

Solution:

(a) By Theorem 7.2 Part (a), $(\frac{3}{13}) \equiv (3)^{(13-1)/2} = 3^6 = 729 \equiv 1$ (mod 13), so $(\frac{3}{13}) = 1$. (Note: This is probably not the easiest way to get this answer, *but* Theorem 7.2 Part (a) is very important for many further results (for example, Theorem 7.2 Parts (b), (c), and (d)! See Problem 7.17.)

(b) Using Theorem 7.4, we note that

$$\{3, 2 \cdot 3, 3 \cdot 3, 4 \cdot 3, 5 \cdot 3, 6 \cdot 3\} = \{3, 6, 9, 12, 15, 18\} \equiv \{3, 6, 9, 12, 2, 5\} \quad (\text{mod } 13)$$

and that there are two values, namely 9 and 12, which exceed 13/2. Hence $(\frac{3}{13}) = (-1)^2 = 1$.

(c) By Theorem 7.5,

$$[3/13] + [6/13] + [9/13] + [12/13] + [15/13] + [18/13]$$

$$= 0 + 0 + 0 + 0 + 1 + 1 = 2,$$

so $(\frac{3}{13}) = (-1)^2 = 1$.

7.7. Use Theorem 7.5 to prove that if p is an odd prime, then the congruence $x^2 \equiv 2 \pmod{p}$ has solutions if and only if $p \equiv 1$ (mod 8) or $p \equiv 7 \pmod 8$.

Solution:

First recall that every odd prime p must be of one of the forms $8k + 1, 8k + 3, 8k + 5$ or $8k + 7$. For the forms $8k + 1$ and $8k + 7$, note that $(p^2 - 1)/8$ will be even while for the forms $8k + 3$ and $8k + 5$, $(p^2 - 1)/8$ will be odd. For example, if $p = 8k + 7$ for some k, then $(p^2 - 1)/8 = (p - 1)(p + 1)/8 = (8k + 6)(8k + 8)/8 = 2(4k + 3)(k + 1)$, which is even; whereas if $p = 8k + 5$, then $(p^2 - 1)/8 = (p - 1)(p + 1)/8 = (8k + 4)(8k + 6)/8 = (2k + 1)(4k + 3)$, which is odd. Using Theorem 7.5, the proof now follows from the fact that $(\frac{2}{p}) = (-1)^{(p^2 - 1)/8}$.

Quadratic Reciprocity

7.8. Evaluate the Legendre symbols (a) $(\frac{71}{73})$, (b)$(\frac{-35}{97})$.

Solution:
(a) There are (at least) two approaches to solving this problem. First, since $71 \equiv -2 \pmod{73}$, we have $(\frac{71}{73}) = (\frac{-1}{73})(\frac{2}{73})$. By Theorem 7.2, $(\frac{-1}{73}) = (-1)^{(73-1)/2} = (-1)^{36} = 1$; by Theorem 7.5, $(\frac{2}{73}) = (-1)^{(73^2-1)/8} = (-1)^{666} = 1$. Hence $(\frac{71}{73}) = 1$. Another approach is to use quadratic reciprocity since both 71 and 73 are prime. By Theorem 7.6, since $(73 - 1)/2$ is even, we know that $(\frac{71}{73}) = (\frac{73}{71}) = (\frac{2}{71})$, Now again applying Theorem 7.5, we have $(\frac{2}{71}) = (-1)^{(71^2-1)/8} = (-1)^{630} = 1$.
(b) Leaving out some details on this part and making use of Theorems 7.2, 7.5, and 7.6, we get (using quadratic reciprocity and the fact that $(97 - 1)/2$ is even)

$$\left(\frac{-35}{97}\right) = \left(\frac{-1}{97}\right)\left(\frac{5}{97}\right)\left(\frac{7}{97}\right) = \left(\frac{-1}{97}\right)\left(\frac{97}{5}\right)\left(\frac{97}{7}\right) = \left(\frac{-1}{97}\right)\left(\frac{2}{5}\right)\left(\frac{6}{7}\right).$$

The three values here are 1, −1 and −1, so $(\frac{-35}{97}) = 1$.

7.9. Use quadratic reciprocity and some of our other results to classify each congruence as solvable or not solvable.
   (a) $x^2 \equiv 11 \pmod{61}$
   (b) $x^2 \equiv 42 \pmod{97}$
   (c) $x^2 \equiv -43 \pmod{79}$

Solution:
(a) Since $(61-1)/2$ is even, we know by quadratic reciprocity that $(\frac{11}{61}) = (\frac{61}{11}) = (\frac{6}{11})$. But 6 is a quadratic non-residue modulo 11, so the given congruence is not solvable.
(b) $(\frac{42}{97}) = (\frac{2}{97})(\frac{3}{97})(\frac{7}{97})$. Since $(97^2 - 1)/8 = 1176$, which is even, $(\frac{2}{97}) = 1$. Since $(97 - 1)/2 = 48$, which is even, by quadratic reciprocity, $(\frac{3}{97}) = (\frac{97}{3}) = (\frac{1}{3}) = 1$. In the same way, $(\frac{7}{97}) = (\frac{97}{7}) = (\frac{-1}{7}) = -1$ because $(7 - 1)/2$ is odd. Hence this congruence is not solvable.

(c) This one is easy. $(\frac{-43}{79}) = (\frac{36}{79}) = 1$ since 36 is a perfect square. Hence this congruence is solvable.

Composite Moduli and the Jacobi Symbol

7.10. Using Definition 7.3, compute the following Jacobi symbols: (a) $(\frac{11}{35})$, (b) $(\frac{19}{35})$, (c) $(\frac{17}{25})$.

Solution:
(a) $(\frac{11}{35}) = (\frac{11}{5})(\frac{11}{7}) = (\frac{1}{5})(\frac{4}{7}) = (1)(1) = 1.$
(b) $(\frac{19}{35}) = (\frac{19}{5})(\frac{19}{7}) = (\frac{4}{5})(\frac{5}{7}) = (1)(-1) = -1.$
(c) $(\frac{17}{25}) = (\frac{17}{5})^2 = (\frac{2}{5})^2 = (-1)^2 = 1.$

7.11. Using Theorem 7.7, classify each of the upper numbers in Problem 7.10 as either a quadratic residue or non-residue modulo the lower number.

Solution:
(a) Since both factors of $(\frac{11}{35})$ are $+1$, 11 is a quadratic residue modulo 35. In fact, $9^2 = 81 \equiv 11 \pmod{35}$.
(b) Since $(\frac{19}{35}) = -1$, 19 is a quadratic non-residue modulo 35.
(c) Despite the fact that $(\frac{17}{25}) = 1$, the presence of even a single factor of -1 says that 17 is quadratic non-residue modulo 25.

7.12. Use generalized quadratic reciprocity to:
(a) determine the value of $(\frac{9}{77})$, and
(b) classify 9 as a quadratic residue or non-residue modulo 77.

Solution:
(a) Because $(9-1)/2$ is even, we know by generalized quadratic reciprocity that $(\frac{9}{77}) = (\frac{77}{9})$. But then $(\frac{77}{9}) = (\frac{5}{9}) = (\frac{5}{3})^2 = (\frac{2}{3})^2 = (-1)^2 = 1$. Hence $(\frac{9}{77}) = 1$.
(b) Because of a -1 in the computation, 9 is a quadratic non-residue modulo 77.

7.8   Supplementary Problems

Quadratic Residues and the Legendre Symbol

7.13. Prove that 5 is a quadratic residue modulo 11 but is a quadratic non-residue modulo 7.

7.14. Find all of the 9 quadratic residues modulo the prime 19.

7.15. Use the definition of the Legendre symbol (Definition 7.2) to find each of the following values:
(a) $\left(\frac{2}{11}\right)$,   (b) $\left(\frac{3}{11}\right)$,   (c) $\left(\frac{-1}{11}\right)$.

7.16. Using Theorem 7.2, find the values of
(a) $\left(\frac{78}{19}\right)$,   (b) $\left(\frac{2}{71}\right)\left(\frac{50}{71}\right)$,   (c) $\left(\frac{46}{47}\right)$.

7.17. Using Theorem 7.2, Part (a), prove Theorem 7.2, Parts (b), (c) and (d).

7.18. Suppose that p is an odd prime and that a and b are both relatively prime to p. Prove that if neither of the congruences $x^2 \equiv a \pmod{p}$ or $x^2 \equiv b \pmod{p}$ has a solution in $\mathbb{Z}_p$, then $x^2 \equiv ab \pmod{p}$ definitely *will* have solutions in $\mathbb{Z}_p$.

Computing the Legendre Symbol

7.19. By Problem 7.2 we know that 3 is a quadratic non-residue modulo 17. Compute the Legendre symbol $\left(\frac{3}{17}\right)$ three different ways presented to us in this chapter:
(a) Use Theorem 7.2 Part (a) to compute $\left(\frac{3}{17}\right)$,
(b) Use Theorem 7.4 (Gauss's Lemma) to compute $\left(\frac{3}{17}\right)$,
(c) Use Theorem 7.5 to compute $\left(\frac{3}{17}\right)$.

7.20. According to Theorem 7.5, if p is an odd prime, then $\left(\frac{2}{p}\right) = (-1)^{(p^2-1)/8}$. This seems to imply that if p is an odd prime, then 8 divides $p^2 - 1$. Prove that this is true.

7.21. Show that if p is an odd prime, then the $\sum_{j=1}^{p-1}\left(\frac{i}{p}\right) = 0$.

Quadratic Reciprocity

7.22. Describe all primes $p > 5$ which have the property that 5 is a quadratic residue modulo p. (Note: See Example 7.7, but this problem is significantly easier than that example once one applies quadratic reciprocity. Why?)

7.23. Using Theorem 7.5 and quadratic reciprocity, classify each congruence as solvable or not solvable.
(a) $x^2 \equiv 10 \pmod{127}$
(b) $x^2 \equiv 71 \pmod{173}$

Composite Moduli and The Jacobi Symbol

7.24. Using Definition 7.3, compute the following Jacobi symbols:
(a) $\left(\frac{11}{35}\right)$, (b) $\left(\frac{19}{45}\right)$, (c) $\left(\frac{17}{27}\right)$.

7.25. Using Theorem 7.7, classify each of the upper numbers in Problem 7.10 as either a quadratic residue or non-residue modulo the lower number.

7.26. Use generalized quadratic reciprocity to:
(a) determine the value of $\left(\frac{27}{55}\right)$, and
(b) classify 27 as a quadratic residue or non-residue modulo 55.

Answers to Selected Supplementary Problems

7.13. $4^2 \equiv 5 \pmod 7$, but the quadratic residues modulo 7 are $\{1, 4, 2\}$.

7.14. $\{1, 4, 9, 16, 6, 17, 11, 7, 5\}$.

7.15. (a) -1, (b) 1, c) -1.

7.16. (a) -1 (See Problem 7.14), (b) 1 (c) -1.

7.19. See Problem 7.6 and its solution.

7.20. Hint: Factor $(p^2 - 1)/2$.

7.21. Suggestion: There are as many residues as non-residues, namely $(p-1)/2$ of each, which can be shown by considering the even and odd powers of a primitive root modulo p.

7.23. Both numbers are non-residues.

7.24. (a) 1,   (b) 1,   (c) −1.

7.25. (a) non-residue,   (b) residue,   (c) non-residue.

7.26. (a) −1,   (b) non-residue.

Chapter 8

Some Fundamental Number Theory Functions

8.1 Introduction

In this chapter we discuss some elementary but important and very useful functions that arise in number theory and in various other areas of mathematics. We have already, in Chapter 5, studied the Euler function $\phi(n)$; and in Chapter 7 we introduced the greatest integer function, which we shall now revisit in the following section. We will then turn to the functions $\tau(n), \sigma(n)$, and $\sigma_k(n)$ which compute, respectively, the number of divisors, the sum of the divisors, and the sum of the k-th powers of the divisors, of the positive integer n. Finally, we will introduce the Möbius function $\mu(n)$ and state and prove the important Möbius Inversion Formula, ending with an application of this formula to the Euler ϕ-function, proving a surprising property of $\phi(n)$. Once again this material is not easy, but hopefully the rewards of encountering and understanding these functions will be well worth the effort.

8.2 The Greatest Integer Function

We begin with the greatest integer function, which was defined in Chapter 7 and used there to study quadratic residues. For completeness, we repeat its definition here.

DOI: 10.1201/9781003193111-8

Definition 8.1. For any real number x, the symbol $[x]$ denotes the greatest integer less than or equal to the value x.

For example, $[6] = 6$, $[4/3] = 1$, and $[-3.5] = -4$.

In the following lemma, we provide a few basic results concerning the greatest integer function.

Lemma 8.1. *Let x and y be real numbers. Then the following properties hold:*
  (a) $[x] \le x < [x] + 1, x - 1 < [x] \le x, 0 \le x - [x] < 1,$
  (b) $[x] = \sum_{1 \le i \le x} 1$ *if $x \ge 0$,*
  (c) $[x + m] = [x] + m$ *if m is an integer,*
  (d) $[x] + [y] \le [x + y] \le [x] + [y] + 1.$

Proof. The first statement of Part (a) is just the definition of $[x]$; the other two statements are simply rearrangements of the first statement.

For Part (b), if $0 \le x < 1$, the sum is empty and hence has value 0. For $x \ge 1$, the sum counts the set of positive integers i which are less than or equal to x, which is exactly the value $[x]$.

Part (c) follows from the definition of $[x]$.

For Part (d) we write $x = n + v, y = m + u$, where n and m are integers and $0 \le v < 1, 0 \le u < 1$. Then we have

$$[x] + [y] = n + m \le [n + v + m + u] = n + m + [v + u] \le$$

$$n + m + 1 = [x] + [y] + 1. \quad \square$$

The next result may look somewhat odd, but it can be used as an alternative way to factor integers of the form $n!$ for any positive integer n.

Theorem 8.2. *If p is a prime and n is a positive integer, then the largest exponent e such that p^e divides $n!$ is*

$$e = \sum_{i=1}^{\infty} [n/p^i],$$

where $[x]$ denotes the greatest integer function.

Before embarking on the proof, let's look at an example.

Example 8.1. Suppose $n = 6$, so that $n! = 6 \cdot 5 \cdot 4 \cdot 3 \cdot 2 \cdot 1 = 720$. Factoring, we obtain $720 = 2^4 \cdot 3^2 \cdot 5$. Let us recover these exponents (4, 2, and 1) using the theorem.

For $p = 5$, $e = \sum_{i=1}^{\infty} [6/5^i] = [6/5] + [6/25] + \cdots = 1 + 0 + \cdots = 1$.

For $p = 3$, $e = \sum_{i=1}^{\infty} [6/3^i] = [6/3] + [6/9] + \cdots = 2 + 0 + \cdots = 2$.

For $p = 2$, $e = \sum_{i=1}^{\infty} [6/2^i] = [6/2] + [6/4] + [6/8] \cdots = 3 + 1 + 0 + \cdots = 4$.

Proof of Theorem 8.2. If $p^i > n$, then $[n/p^i] = 0$, so the summation is eventually all 0's. Thus it is a finite sum which can now be proved by mathematical induction. (See Appendix A.)

For the base case $n = 1$ and for any prime p, $[1/p^i] = 0$ for all $i \geq 1$, so the result holds. Now for the inductive step, we assume that for $n > 1$ the theorem is true for $(n-1)!$, and we must show that this implies that it also holds for $n!$. Let k be the largest integer so that p^k divides n. Since $n! = n \cdot (n-1)!$, we must show that $\sum [n/p^i] - \sum [(n-1)/p^i] = k$. We note that the difference $[n/p^i] - [(n-1)/p^i]$ equals 1 if p^i divides n, and equals 0 if p^i does not divide n. We claim then that

$$\sum_{i=1}^{\infty} [n/p^i] - \sum_{i=1}^{\infty} [(n-1)/p^i] = \sum_{i=1}^{\infty} ([n/p^i] - [(n-1)/p^i]) = k.$$

If p does not divide n, then all quantities above are 0, so the equality holds. If p does divide n, then the sum of the difference above is $\sum_1^k 1 = k$, and again the equality holds, and so the proof is complete. $\square$

Example 8.2. Following up on Example 8.1 with $n = 6$ and $p = 2$, we see that k in the proof above equals 1 (i.e., 2 divides 6 but 4 does not), and we have

$$\sum ([6/2^i] - [5/2^i]) = ([6/2] - [5/2]) + ([6/4] - [5/4]) = (3-2) + (1-1)$$
$$= 1 + 0 = 1,$$

as expected.

We now provide one application of the previous theorem.

Corollary 8.3. *If $a_i \geq 0$ for $i = 1, \ldots, r$ and $a_1 + \cdots + a_r = n$ then the expression*

$$\frac{n!}{a_1! \cdots a_r!}$$

is an integer.

Proof. Our proof will consist of showing that every prime which divides the numerator occurs to at least as high a power as it occurs in the denominator. Let p be such a prime and suppose the highest power of p dividing $n!$ is e and that for each j, the highest power of p dividing $a_j!$ is e_j. We must show then that $e \geq e_1 + e_2 + \cdots + e_r$. But by Theorem 8.2, we know that $e = \sum [n/p^i]$ and for each j, $e_j = \sum [a_j/p^i]$. Using Part (d) of Lemma 8.1, we have

$$[a_1/p^i] + \cdots + [a_r/p^i] \leq [(a_1 + \cdots + a_r)/p^i] = [n/p^i].$$

We now sum over the powers i of p, obtaining

$$\sum_{i=1}^{\infty} [n/p^i] \geq \sum_{i=1}^{\infty} [a_1/p^i] + \cdots + \sum_{i=1}^{\infty} [a_r/p^i],$$

which completes the proof. $\square$

Example 8.3. The binomial coefficient $\binom{n}{k}$ has the value $\frac{n!}{k!(n-k)!}$. Corollary 8.3 guarantees that this number is an integer. For example, consider the binomial coefficient $\binom{12}{4} = \frac{12!}{4!8!} = 495$. Note that $12!$ is divisible by 2, 3, 5, 7, and 11. Let's work through each of these:

Let $p = 2$. Using Theorem 8.2 and a variation of the notation there (e.g., e_{12} for e, etc.), we have $e_{12} = [12/2] + [12/4] + [12/8] = 10$, $e_4 = [4/2] + [4/4] = 3$, and $e_8 = [8/2] + [8/4] + [8/8] = 7$. Since $10 = 3 + 7$, there are no "extra" 2's in $12!$, and hence no 2's in the answer.

Let $p = 3$. We have $e_{12} = [12/3] + [12/9] = 4 + 1 = 5$, $e_4 = [4/3] = 1$, and $e_8 = [8/3] = 2$, so the answer has two 3's, i.e., a factor of 9.

Let $p = 5$. We have $e_{12} = [12/5] = 2$, $e_4 = 0$, and $e_8 = [8/3] = 1$, so we have one of 5 in the answer.

Let $p = 7$. We have $e_{12} = [12/7] = 1$, $e_4 = 0$, and $e_8 = [8/7] = 1$, so we have no 7's in the answer.

Let $p = 11$. We have $e_{12} = [12/11] = 1$ and $e_4 = e_8 = 0$, so we have one factor of 11 in the answer.

We confirm then that $\binom{12}{4} = 9 \cdot 5 \cdot 11 = 495$. (Note: This is certainly not the easiest method to get this answer, but using it was intended to illustrate Corollary 8.3 and its proof using Theorem 8.2.)

8.3 The Functions $\tau(n)$, $\sigma(n)$, $\sigma_k(n)$

Functions like the Euler ("phi") function $\phi(n)$ (introduced in Chapter 5) which are just defined for positive integers are called *numerical functions*. We will now consider several other such functions. We note that the greatest integer function in the previous section is *not* a numerical function as defined here since it is rather a function from the real numbers to the integers.

We make the following definitions, where the symbol τ is the Greek letter "tau" and the symbol σ is the Greek letter "sigma."

Definition 8.2. Let n be a positive integer. Then

1. $\tau(n)$ denotes the number of divisors of n (including 1 and n itself).

2. $\sigma(n)$ denotes the sum of the divisors of n.

3. $\sigma_k(n)$ denotes the sum of the k-th powers of the divisors of n.

Example 8.4.

1. $\tau(6) = 4$ since the positive integers $1, 2, 3$, and 6 are the divisors of 6.

2. $\sigma(6) = 1 + 2 + 3 + 6 = 12$. (Note that 6 is a so-called "perfect number" since the sum of its *proper* divisors (i.e., 1, 2, and 3) is itself. In general a perfect number n satisfies that $\sigma(n) = 2n$. Are there other perfect numbers? See Open Problem 11.2.)

3. $\sigma_2(6) = 1^2 + 2^2 + 3^2 + 6^2 = 50$.

We note that the functions $\tau(n)$ and $\sigma(n)$ are actually both special cases of the function $\sigma_k(n)$ since $\tau(n) = \sigma_0(n)$ and $\sigma(n) = \sigma_1(n)$.

We now introduce a convenient notation. If f is a numerical function, we write $\sum_{d|n} f(d)$ and $\prod_{d|n} f(d)$ for the sum, and product, of the function f taken over all positive divisors d of n. As a result of this notation, we have

$$\tau(n) = \sum_{d|n} 1, \quad \sigma(n) = \sum_{d|n} d, \quad \sigma_k(n) = \sum_{d|n} d^k.$$

Theorem 8.4. *If $n = p_1^{e_1} \cdots p_r^{e_r}$ is the prime factorization of the positive integer n, then*

$$\tau(n) = (e_1 + 1) \cdots (e_r + 1)$$

while $\tau(1) = 1$.

Proof. A positive integer d divides the integer n if and only if $d = p_1^{k_1} \cdots p_r^{k_r}$ where $0 \le k_i \le e_i$ for each $i = 1, \ldots, r$. Thus the total number of divisors d of n is given by the product $(e_1+1) \cdots (e_r+1)$. $\square$

It follows from Theorem 8.4 that if $\gcd(m, n) = 1$, then, just as is the case for the Euler ϕ-function, $\tau(mn) = \tau(m)\tau(n)$. Hence these two functions are examples of an important class of numerical functions which we now define.

Definition 8.3. If f is a numerical function and $f(mn) = f(m)f(n)$ for every pair of positive integers m and n which are relatively prime, then the function f is *multiplicative*. If $f(mn) = f(m)f(n)$ independent of whether m and n are relatively prime, then the function f is *totally multiplicative*.

Example 8.5. Is the function $\sigma(n)$, like $\tau(n)$, multiplicative? Let's look at one example to get a hint. In doing this we are also previewing the proof method used in Theorem 8.5 below. So, let's examine $\sigma(60)$. Since $60 = 2^2 \cdot 3 \cdot 5$, we know from Theorem 8.4

that $\tau(60) = (2+1)(1+1)(1+1) = 12$; i.e., 60 has 12 divisors. We have $\sigma(4) = 1+2+4$, $\sigma(15) = 1+3+5+15$, and note that 4 and 15 are relatively prime. But now

$$\sigma(4)\sigma(15) = (1+2+4)(1+3+5+15) =$$

$$(1+3+5+15) + (2+6+10+30) + (4+12+20+60).$$

Here then are the 12 divisors of 60 (which, by the way, add up to 168), and so we see that $\sigma(60) = \sigma(4)\sigma(15)$. Since this is a single example, it does not, of course, prove that the function σ is multiplicative, but it gives us some evidence that it may be.

The following very general and important result tells us what we want to know about $\sigma(n)$ among many other numerical functions.

Theorem 8.5. *Let $f(n)$ be a multiplicative function and let $F(n) = \sum_{d\mid n} f(d)$. Then the function $F(n)$ is also multiplicative.*

Proof. Assume the integers m and n are relatively prime. If $n = p_1^{\alpha_1} \cdots p_r^{\alpha_r}$ and $m = q_1^{\beta_1} \cdots q_s^{\beta_s}$ with positive exponents α_i and β_j, then $p_1, \ldots, p_r, q_1, \ldots, q_s$ are all *distinct* primes. Hence the positive divisors d_1 of n are the numbers $d_1 = p_1^{\gamma_1} \cdots p_r^{\gamma_r}$ for all possible choices $0 \le \gamma_i \le \alpha_i$. Similarly the positive divisors d_2 of m are the numbers $d_2 = q_1^{\delta_1} \cdots q_s^{\delta_s}$ for all possible choices $0 \le \delta_j \le \beta_j$. Thus as d_1 runs through the divisors of n and d_2 runs through the divisors of m, their product $d_1 d_2$ runs through the integers

$$d = d_1 d_2 = p_1^{\gamma_1} \cdots p_r^{\gamma_r} q_1^{\delta_1} \cdots q_s^{\delta_s}, 0 \le \gamma_i \le \alpha_i, 0 \le \delta_j \le \beta_j,$$

which are the positive divisors of $nm = p_1^{\alpha_1} \cdots p_r^{\alpha_r} q_1^{\beta_1} \cdots q_s^{\beta_s}$.
We thus have

$$\sum_{d_1\mid n} \sum_{d_2\mid m} f(d_1 d_2) = \sum_{d\mid nm} f(d).$$

Since $\gcd(d_1, d_2) = 1$, it follows that

$$F(nm) = \sum_{d|nm} f(d) = \sum_{d_1|n}\sum_{d_2|m} f(d_1 d_2) = \sum_{d_1|n}\sum_{d_2|m} f(d_1)f(d_2)$$

$$= \sum_{d_1|n} f(d_1) \sum_{d_2|m} f(d_2) = F(n)F(m),$$

so the proof is complete. □

Corollary 8.6. *For every non-negative integer k, the function $\sigma_k(n)$ is multiplicative. Hence, in particular, $\tau(n) = \sigma_0(n)$ and $\sigma(n) = \sigma_1(n)$ are multiplicative.*

Proof. Let the function $f(n)$ in Theorem 8.5 be $f(n) = n^k$. Since $n^k m^k = (nm)^k$, $f(n)$ is multiplicative, and so by that theorem $F(n) = \sigma_k(n)$ is multiplicative. □

We finish this section with a result on computing $\sigma(n)$ which is parallel to Theorem 8.4 on computing $\tau(n)$.

Theorem 8.7. *If $n = p_1^{e_1} \cdots p_r^{e_r}$ is the prime factorization of the positive integer n, then*

$$\sigma(n) = \prod_{i=1}^{r} \frac{p_i^{e_i+1} - 1}{p_i - 1},$$

and $\sigma(1) = 1$.

Proof. Since $\sigma(n)$ is multiplicative, we have $\sigma(n) = \prod_{i=1}^{r} \sigma(p_i^{e_i})$. Because the divisors of $p_i^{e_i}$ are $1, p_i, p_i^2, \ldots, p_i^{e_i}$, we can use the "sum of a geometric series formula" (i.e., $\sum_{i=0}^{e} x^i = (x^{(e+1)} - 1)/(x-1)$) to get a sum of $(p_i^{e_i+1} - 1)/(p_i - 1)$ for each prime factor p_i, and multiplying all these together completes the proof. □

Example 8.6. Following up on Example 8.5, we can use Theorem 8.7 to compute $\sigma(60)$. Again, since $60 = 2^2 \cdot 3 \cdot 5$, we get

$$\sigma(60) = \frac{2^3 - 1}{2 - 1} \cdot \frac{3^2 - 1}{3 - 1} \cdot \frac{5^2 - 1}{5 - 1} = 7 \cdot 4 \cdot 6 = 168,$$

as we saw directly in Example 8.5.

8.4 The Möbius Inversion Formula

We now study another important example of a numerical function which turns out to be multiplicative and which is named after August Möbius (1790 - 1868). It is denoted by μ, which is the Greek letter "mu."

Definition 8.4. The Möbius function $\mu(n)$ is defined by $\mu(n) = 1$ if $n = 1$, 0 if $a^2|n$ for some integer $a > 1$, and $(-1)^r$ if $n = p_1 \cdots p_r$, where the p_i are distinct primes.

Example 8.7. Let's compute μ for the first ten positive integers: $\mu(1) = 1$, $\mu(2) = -1$, $\mu(3) = -1$, $\mu(4) = 0$, $\mu(5) = -1$, $\mu(6) = 1$, $\mu(7) = -1$, $\mu(8) = 0$, $\mu(9) = 0$, and $\mu(10) = 1$.

Here are the first needed properties of the Möbius function.

Theorem 8.8. *The Möbius function $\mu(n)$ is multiplicative and, in addition,*

$$\sum_{d|n} \mu(d) = \begin{cases} 1 & \text{if } n = 1 \\ 0 & \text{if } n > 1. \end{cases}$$

Proof. Suppose $n = p_1 \cdots p_r$ and $m = q_1 \cdots q_s$ where all the primes p_i and q_j are distinct (i.e., n and m are both "square-free" and $\gcd(n, m) = 1$). Then $\mu(nm) = (-1)^{r+s} = (-1)^r(-1)^s = \mu(n)\mu(m)$, so the Möbius function is multiplicative.

For the second statement, if $F(n) = \sum_{d|n} \mu(d)$, then $F(n)$ is multiplicative by Theorem 8.5. We have $F(1) = \mu(1) = 1$; and, for each prime divisor p of $n > 1$, we have

$$F(p^e) = \sum_{k=0}^{e} \mu(p^k) = \mu(1) + \mu(p) + \mu(p^2) + \cdots + \mu(p^e)$$

$$= 1 + (-1) + 0 + \cdots + 0 = 0,$$

so we are done. $\square$

We are now able to state and prove a very useful formula. We know that given a numerical function $f(n)$, we can use it to compute the corresponding "summation function" $F(n) = \sum_{d|n} f(d)$. This theorem says that, on the other hand, if you have certain values of the summation function, you can use these and the Möbius function to "recover" the original function $f(n)$. Here is the theorem:

Theorem 8.9. (*Möbius Inversion Formula*) *Let $f(n)$ be a given numerical function. If $F(n) = \sum_{d|n} f(d)$ for every positive integer n, then*

$$f(n) = \sum_{d|n} \mu(d) F(n/d).$$

Proof. The argument contains four equalities, the second of which may be a bit tricky to follow. In that equality the two sums get multiplied together, creating a single doubly-subscripted sum. The switch of δ with d in the subscripting is simply a rearrangement of that sum. We supply a brief justification for each equality:

$$\sum_{d|n} \mu(d) F(n/d) = \sum_{d|n} \mu(d) \sum_{\delta|(n/d)} f(\delta) \quad \text{(Definition of } F(n/d))$$

$$= \sum_{\delta|n} \sum_{d|(n/\delta)} \mu(d) f(\delta) \quad \text{(See comment above)}$$

$$= \sum_{\delta|n} f(\delta) \sum_{d|(n/\delta)} \mu(d) \quad \text{(Factor out each } f(\delta))$$

$$= f(n) \quad \text{(By Theorem 8.8, } \sum_{d|n/\delta} \mu(d) = 0 \text{ unless } \delta = n). \quad \square$$

Here is an illustration of Möbius Inversion.

Example 8.8. We use the function $\sigma(n)$ to illustrate the theorem; specifically, we examine the cases of $n = 10$ and $n = 9$, which are somewhat different since 10 is "square-free" but 9 is not.

We first set $n = 10$ and note that $\sigma(10) = 1 + 2 + 5 + 10 = 18$. On the right-hand side of the theorem's equation, where for any positive integer m, $F(m) = \sum_{d|m} \sigma(d)$, we need the values $F(1) = \sigma(1) = 1$, $F(2) = \sigma(1)+\sigma(2) = 1+3 = 4$, $F(5) = \sigma(1)+\sigma(5) = 1+6 = 7$, and $F(10) = \sigma(1)+\sigma(2)+\sigma(5)+\sigma(10) = 1+3+6+18 = 28$. Here then is that right-hand side:

$$\sum_{d|10} \mu(d)F(10/d) = \mu(1)F(10) + \mu(2)F(5) + \mu(5)F(2) + \mu(10)F(1)$$

$$= (1)(28) + (-1)(7) + (-1)(4) + (1)(1) = 18,$$

as guaranteed by the theorem.

Now set $n = 9$, and we have $\sigma(9) = 1 + 3 + 9 = 13$. We need $F(1) = 1$, $F(3) = \sigma(1) + \sigma(3) = 1 + 4 = 5$, and $F(9) = \sigma(1) + \sigma(3) + \sigma(9) = 1 + 4 + 13 = 18$. This time the theorem's right-hand side is

$$\sum_{d|9} \mu(d)F(9/d) = \mu(1)F(9) + \mu(3)F(3) + \mu(9)F(1)$$

$$= (1)(18) + (-1)(5) + (0)(1) = 13,$$

again as guaranteed by the theorem.

Möbius Inversion (Theorem 8.9) is actually an "if and only if" result, as is shown by the following theorem, which we state without proof.

Theorem 8.10. *If $f(n) = \sum_{d|n} \mu(d)F(n/d)$ for every positive integer n, then $F(n) = \sum_{d|n} f(d)$.*

We point out that in the last two theorems, neither the function $f(n)$ nor the function $F(n)$ is required to be multiplicative.

Let's finish this chapter with a nice application of Theorem 8.10. The question we wish to answer is, If $\phi(n)$ is the Euler function (which, recall, counts how many of the numbers between 1 and n are relatively prime to n), what is $\sum_{d|n} \phi(d)$? It's always a good idea to look at a couple of examples:

$$\sum_{d|12} \phi(d) = \phi(1) + \phi(2) + \phi(3) + \phi(4) + \phi(6) + \phi(12)$$

$$= 1 + 1 + 2 + 2 + 2 + 4 = 12,$$

$$\sum_{d|27} \phi(d) = \phi(1) + \phi(3) + \phi(9) + \phi(27) = 1 + 2 + 6 + 18 = 27.$$

This data suggests the following result:

Theorem 8.11. *If $\phi(n)$ is the Euler function, then*

$$\sum_{d|n} \phi(d) = n.$$

Proof. Using the notation of Theorem 8.10, we let $F(n) = n$. We now compute the function $f(n)$ defined by the equation in that theorem. Because F is a multiplicative function, it suffices to compute $f(p^e)$ for a prime power p^e. We have then

$$f(p^e) = \sum_{d|p^e} \mu(d)(p^e/d) = \mu(1)(p^e/1)$$

$$+\mu(p)(p^e/p)) + \mu(p^2)(p^e/p^2) + \cdots + \mu(p^e)(p^e/p^e)$$

$$= (1)(p^e)+(-1)(p^{e-1})+(0)(p^{e-2})+\cdots+(0)(1) = p^e - p^{e-1} = \phi(p^e).$$

Thus, by multiplicativity, $f(n) = \phi(n)$ for all positive integers n. But now by Theorem 8.10,

$$n = F(n) = \sum_{d|n} f(d) = \sum_{d|n} \phi(d),$$

and we are done. □

In Chapter 10 we will see another very nice application of Möbius Inversion.

8.5 Summary

Here is a summary of the functions we have studied in this chapter. The first is defined on all real numbers; all the others are defined on all positive integers.

Name	Definition
greatest integer function $[x]$	greatest integer $\leq x$
Euler function $\phi(n)$	counts $1 \leq m \leq n$ with $\gcd(m,n) = 1$
tau function $\tau(n)$	counts positive divisors of n
sigma function $\sigma(n)$	sum of positive divisors of n
sigma power function $\sigma_k(n)$	sum of k-th powers of positive divisors of n 1 if $n = 1$
Möbius function $\mu(n)$	$(-1)^r$ if $n = p_1 p_2 \cdots p_r$, all p_i distinct primes 0 otherwise

All of these functions have numerous applications in number theory and across various other areas of mathematics. We have already seen (in Chapters 5 and 6) the Euler ϕ-function "in action," and the same (in the previous and this chapter) for the greatest integer function. The Möbius function $\mu(n)$ and Möbius Inversion were used in this chapter to get information about the Euler function. Also, in Chapter 10 the Möbius function and formula will be used to count "monic irreducible polynomials" of degree n over the "finite field $\mathbb{F}_q$" (where q is a prime power). Finally, in Chapter 11 we will introduce and discuss a new numerical function $\pi(n)$ (not to be confused with the famous real number π) which counts the number of prime numbers between 2 and n.

8.6 Solved Problems

The Greatest Integer Function

8.1. (a) Use Theorem 8.2 to find the highest power of 2 in 20!.
(b) Do the same for the highest power of 3 in 20!.
(c) Do the same for the highest power of 5 in 20!.
(d) Do the same for the highest power of 7 in 20!.

Solution:
(a) $\sum_{i=1}^{\infty} [20/2^i] = [20/2] + [20/4] + [20/8] + [20/16] = 10 + 5 + 2 + 1 = 18$.
(b) In the same way, $[20/3] + [20/9] = 6 + 2 = 8$.
(c) In the same way, $[20/5] = 4$.

(d) In the same way, $[20/7] = 2$.

8.2. (a) As in Example 8.3, compute the binomial coefficient $\binom{10}{7}$ using its definition.

(b) Use Theorem 8.2 to compute $\binom{10}{7}$ by counting the number of factors of 2, 3, 5, and 7 in each of $10!$, $7!$ and $3!$.

Solution:

(a) $\binom{10}{7} = 10!/(7!3!) = (10 \cdot 9 \cdot 8)/(3 \cdot 2 \cdot 1) = 120$.

(b) Using the notation of the example, we have:

For $p = 2$, $e_{10} = [10/2] + [10/4] + [10/8] = 5 + 2 + 1 = 8$, $e_7 = [7/2]+[7/4] = 3+1 = 4$ and $e_3 = [3/2] = 1$. Since $8-4-1 = 3$, the answer has three factors of 2.

For $p = 3$, $e_{10} = [10/3] + [10/9] = 3 + 1 = 4$, $e_7 = [7/3] = 2$ and $e_3 = [3/3] = 1$. Since $4 - 2 - 1 = 1$, the answer has one factor of 3.

For $p = 5$, $e_{10} = [10/5] = 2$, $e_7 = [7/5] = 1$ and $e_3 = [3/5] = 0$. Since $2 - 1 = 1$, the answer has one factor of 5.

For $p = 7$, $e_{10} = [10/7] = 1$, $e_7 = [7/7] = 1$ and $e_3 = [3/7] = 0$. Since $1 - 1 = 0$, the answer has no factors of 7.

We conclude that $\binom{10}{7} = 2^3 \cdot 3 \cdot 5 = 120$, as expected.

8.3. Find formulas for the highest power e of the prime p so that p^e divides

(a) the product of the $2 \cdot 4 \cdot 6 \cdots (2n)$ of the first n even integers,

(6) the product of the first n odd integers.

Solution:

(a) Factoring 2 out of each of the n factors, we have $2 \cdot 4 \cdot 6 \cdots (2n) = 2^n n!$, so by Theorem 8.2, for odd p we get $e = \sum_{i=1}^{\infty}[n/p^i]$, and for $p = 2$ we get $e = n + \sum_{i=1}^{\infty}[n/2^i]$.

(b) By rearranging $(2n)!$, we see that

$$(2n)! = 1 \cdot 2 \cdots (2n) = (1 \cdot 3 \cdots (2n - 1)) \cdot (2 \cdot 4 \cdots (2n)),$$

so we can get the answers from Theorem 8.2 applied to $(2n)!$ and from Part (a). Hence, for odd p we get $e = \sum_{i=1}^{\infty}[2n/p^i] - [n/p^i]$, and for $p = 2$ we get, of course, 0.

The Functions $\phi(n)$, $\tau(n)$, $\sigma(n)$, and $\sigma_k(n)$

8.4. Find the value of:
(a) $\phi(20)$,   (b) $\tau(20)$,   (c) $\sigma(20)$,   (d) $\sigma_2(20)$.

Solution:
(a) As we learned in Chapter 5, $\phi(20) = \phi(2^2)\phi(5) = 2 \cdot 4 = 8$.
(b) Since the divisors of 20 are 1, 2, 4, 5, 10, and 20, $\tau(20) = 6$.
(c) $\sigma(20) = 1 + 2 + 4 + 5 + 10 + 20 = 42$.
(d) $\sigma_2(20) = 1^2+2^2+4^2+5^2+10^2+20^2 = 1+4+16+25+100+400 = 546$.

8.5. (a) Using Theorem 8.4, compute $\tau(1,000)$.
(b) Using Theorem 8.7, compute $\sigma(1,000)$.

Solution:
(a) Since $1,000 = 10^3 = 2^3 5^3$, $\tau(1,000) = (3 + 1)(3 + 1) = 16$.
(b) $(2^4-1)/(2-1) \cdot (5^4-1)/(5-1) = 15 \cdot (624/4) = 15 \cdot 156 = 2,340$.

8.6. Find the smallest positive integer m so that $\sigma(n) = m$ has (a) no solutions, (b) exactly one solution, (c) exactly two solutions.

Solution:
(a) Since $\sigma(1) = 1$ and $\sigma(2) = 3$, we see that $m = 2$ cannot occur as a value of σ.
(b) Since $\sigma(1) = 1$ but for $n \geq 2$ we have $\sigma(n) > 2$, $m = 1$ is the smallest σ value with this property.
(c) This one is trickier. We have $\sigma(3) = 4$, $\sigma(4) = 7$, $\sigma(5) = 6$ and $\sigma(6) = 12$. No integer $n > 6$ can have a σ value of 4, 7, or 6, *but* note that $\sigma(11) = 12$. Moreover, none of $\sigma(7)$ through $\sigma(10)$ have the value 12, and for $n \geq 12$, $\sigma(n) > 12$. Hence $m = 12$ is the smallest σ value to occur exactly twice.

The Möbius Function and Möbius Inversion

8.7. Calculate the value of the Möbius function $\mu(n)$ for each of the following numbers: (a) $n = 7$,   (b) $n = 49$,   (c) $n = 35$,   (d) $n = 105$,   (e) $n = 525$.

Solution:
(a) $\mu(7) = -1$,   (b) $\mu(7^2) = 0$,   (c) $\mu(5 \cdot 7) = (-1)^2 = 1$,
(d) $\mu(3 \cdot 5 \cdot 7) = (-1)^3 = -1$,   (e) $\mu(3 \cdot 5^2 \cdot 7) = 0$.

8.8. Evaluate the sum $\sum_{j=1}^{\infty} \mu(j!)$.

Solution:
Note that for all $j \geq 4$, $j!$ contains a factor of 2^2, so $\mu(j) = 0$.
Hence the sum is $\mu(1!) + \mu(2!) + \mu(3!) = \mu(1) + \mu(2) + \mu(3 \cdot 2) = 1 + (-1) + 1 = 1$.

8.9. As was done in Example 8.8 for the sigma function, verify what the Möbius Inversion Formula (Theorem 8.9) says for the tau function applied to the case $n = 10$.

Solution:
We have $\tau(10) = 4$, and we wish to "recover" this value from the inversion formula. Letting $F(n) = \sum_{d|n} \tau(d)$, we need the values $F(1) = 1$, $F(2) = 1 + 2 = 3$, $F(5) = 1 + 2 = 3$ and $F(10) = 1 + 2 + 2 + 4 = 9$. The inversion formula is now

$$\sum_{d|10} \mu(d)\tau(10/d) = (1)(9) + (-1)(3) + (-1)(3) + (1)(1) = 4$$

as expected.

8.10. Assume that the canonical factorization of the integer n is $p_1^{k_1} p_2^{k_2} \cdots p_r^{k_r}$. Show that $\sum_{d|n} \mu(d)\tau(d) = (-1)^r$.

Solution:
Since both μ and τ are multiplicative, it suffices to do the computation for each $p_i^{k_i}$. We have then

$$\sum_{d|p_i^{k_i}} \mu(d)\tau(d) = \mu(1)\tau(1) + \mu(p_i)\tau(p_i) + \mu(p_i^2)\tau(p_i^2) + \cdots + \mu(p_i^{k_i})\tau(p_i^{k_i})$$

$$= (1)(1) + (-1)(2) + (0)(3) + \cdots + (0)(k_i + 1) = -1.$$

Multiplying out, we get the desired result.

8.7 Supplementary Problems

The Greatest Integer Function

8.11. (a) Use Theorem 8.2 to find the highest power of 2 in 36!.
(b) Do the same for the highest power of 3 in 36!.
(c) Do the same for the highest power of 5 in 36!.
(d) Do the same for the highest power of 11 in 36!.

8.12. If 100! were multiplied out, how many zeros would there be at the end?

8.13. (a) As in Example 8.3, compute the binomial coefficient $\binom{10}{6}$ using its definition.
(b) Use Theorem 8.2 to compute $\binom{10}{6}$ by counting the number of factors of 2, 3, 5 and 7 in each of 10!, 6! and 4!.

8.14. For what real numbers x is it true that
    (a) $[x + 2] = [x] + 2$,
    (b) $[2x] = 2[x]$.
(Suggestion: It's very often a good idea to simply compute some examples and try to recognize a pattern. For example, for Part (b), note that $2[1.4] = 2 \cdot 1 = 2 = [2.8]$, but $2[1.6] = 2 \cdot 1 = 2 \neq [3.2]$.)

The Functions $\phi(n)$, $\tau(n)$, $\sigma(n)$, and $\sigma_k(n)$

8.15. Find the value of:
(a) $\phi(30)$,   (b) $\tau(30)$,   (c) $\sigma(30)$,   (d) $\sigma_2(30)$.

8.16. Find the smallest positive integer n so that
    (a) $\phi(n) = 6$,
    (b) $\tau(n) = 6$,
    (c) $\sigma(n) = 6$.

8.17. (a) Using Theorem 8.4, compute $\tau(864)$.
(b) Using Theorem 8.7, compute $\sigma(864)$.

8.18. Using Theorem 8.7 as a model, find a formula for the sum $\sigma_k(n)$, the sum of the k-th powers of the divisors of the positive integer n.

8.19. Given a positive integer $m > 1$, show there are infinitely many positive integers n satisfying $\tau(n) = m$.

8.20. By Theorem 8.5, we know that if $f(n)$ is multiplicative, then $F(n) = \sum_{d|n} f(d)$ is also multiplicative. One might conjecture then that if $f(n)$ is *totally* multiplicative (i.e., $f(mn) = f(n)f(m)$ even if n and m are not relatively prime), then $F(n)$ would also be totally multiplicative. Show that this conjecture is false, using the example of the totally multiplicative function $f(n) = n$.

The Möbius Function and Möbius Inversion

8.21. Calculate the value of the Möbius function $\mu(n)$ for each of the following numbers: (a) $n = 11$,   (b) $n = 121$,   (c) $n = 22$, (d) $n = 66$,   (e) $n = 132$.

8.22. Similar to what was done in Example 8.8 for the sigma function, verify what the Möbius Inversion Formula (Theorem 8.9) says for the tau function applied to the case $n = 12$. We know that $\tau(12) = 6$, but now recover that value from the inversion formula.

8.23. Assume that the canonical factorization of the integer n is $p_1^{k_1} p_2^{k_2} \cdots p_r^{k_r}$. Show that $\sum_{d|n} \mu(d)\sigma(d) = (-1)^r p_1 p_2 \cdots p_r$. (Hint: Both μ and σ are multiplicative.)

8.24. Find the smallest positive integer n with the property that $\mu(n) + \mu(n+1) + \mu(n+2) = 3$. (Hint: This will happen if and only if all three of n, $n+1$, and $n+2$ are the product of an even number of distinct primes. The answer is less than 40.)

Answers to Selected Supplementary Problems

8.11. (a) 34,   (b) 17,   (c) 8,   (d) 3.

8.12. The highest power of 5 in 100! will determine this. That power is 24.

8.13. (a) 210.

(b) Factors of 2 in the quotient: $8 - (4 + 3) = 1$, factors of 3: $4 - (2 + 1) - 1$, factors of 5: $2 - (1 + 0) = 1$, and factors of 7: $1 - (0 + 0) = 1$. Hence the answer is $2 \cdot 3 \cdot 5 \cdot 7 = 210$, as expected.

8.14. (a) x can be any real number.
(b) If $x \geq 0$, its decimal part must be less than .5. If $x < 0$, its decimal part must be greater than .5.

8.15. (a) 8, (b) 8, (c) 72, (d) 1300.

8.16. (a) 7, (b) 12, (c) 5.

8.17. (a) 24, (b) 2,520.

8.18. $\sigma_k(p_1^{e_1} \cdots p_r^{e_r}) = \prod_{i=1}^{r}(p_i^{k(e_i+1)} - 1)/(p_i^k - 1)$.

8.19. Let $n = p^{m-1}$ with p any prime number.

8.21. (a) -1, (b) 0, (c) 1, (d) -1, (e) 0.

8.24. $n = 33$.

Chapter 9

Diophantine Equations

9.1   Introduction

In this chapter we shall study equations having two or more unknowns in which all coefficients are integers and in which we are
seeking integer solutions. Such equations are called *Diophantine
equations*, named after Diophantus of Alexandria (AD 201-215 -
AD 285-299). Just as is true with the "ordinary" equations with
real number coefficients which we study in basic algebra, there is
no general method to solve Diophantine equations. For example,
in algebra we try some method to solve linear equations, some
different method to solve quadratic equations, and so on. This is,
as we shall see, also the case when trying to solve Diophantine
equations of different sorts. This topic can be very broad, but we
shall limit our discussion to a few specific types of Diophantine
equations.

9.2   The Linear Equation $ax + by = c$

We first consider the simplest Diophantine equation, namely a
linear equation where we seek integers x and y so that for given
integers a, b, c, we have $ax + by = c$. The problem is rather trivial
if one of a or b is 0. For example if $a = 0$, then the equation $by = c$
has an integer solution if and only if b divides c. Hence we assume

DOI: 10.1201/9781003193111-9

that neither a nor b is zero. If $d = \gcd(a, b)$, the greatest common divisor of the integers a and b, then the above linear equation has no solution if d does not divide c since d clearly *does* divide $ax+by$.

Suppose then that $d = \gcd(a, b)$ divides c. We can now use the following procedure to first find an initial solution to our equation and then find out how to generate infinitely many solutions from that initial one:

(1) Since d divides c, we can divide our equation through by d, and the resulting equation will have the exact same solutions as the original equation. For simplicity, we shall denote the new reduced equation by the symbols $ax + by = c$, where now a and b are relatively prime.

(2) Use either trial and error or the Euclidean Algorithm to find a solution (x_0, y_0) of the equation $ax+by = 1$. For a reminder of how to find (x_0, y_0) using the Euclidean Algorithm, see the discussion following Example 1.5 as well as Solved Problems 1.10 and 1.11.

(3) Multiplying both sides of the solved equation in (2) by c, we obtain the equation $a(cx_0) + b(cy_0) = c$, and so we have found our initial solution $(x_1 = cx_0, y_1 = cy_0)$ of the given equation.

(4) We can now use the solution (x_1, y_1) to generate infinitely many solutions, basically by having the x solution go up and the y solution go down, or vice versa. Letting t be any integer, we claim that $(x_1 + bt, y_1 - at)$ is also a solution. Checking this:

$$a(x_1 + bt) + b(y_1 - at) = ax_1 + abt + by_1 - abt = ax_1 + by_1 = c.$$

Since t can range over all integers, we obtain infinitely many solutions of our equation.

Here we illustrate this procedure.

Example 9.1. (a) Consider first the linear Diophantine equation $6x + 12y = 8$. Since $\gcd(6, 12) = 6$ does not divide 8, we know that this equation has no integer solutions.
(b) Now consider the equation $6x+10y = 8$. Since $d = \gcd(6, 10) = 2$ does divide 8, we can use our procedure to find its infinitely many solutions.

(1) We divide out the common factor of 2 to create a new, equivalent equation $3x + 5y = 4$ whose solutions are identical to those of our original equation.

(2) We first seek a solution of the equation $3x + 5y = 1$. By quick trial and error, we see that $(2, -1)$ is a solution. (If a solution of the current equation cannot be found quickly by trial and error, we can use the Euclidean Algorithm.)

(3) Multiplying through by $c = 4$, we get $3(8) + 5(-4) = 4$, so $(8, -4)$ is our initial solution.

(4) Hence the general solution is $(8 + 5t, -4 - 3t)$ for any integer t. For example, if $t = 2$, we get $(18, -10)$, and we check: $3(18) + 5(-10) = 54 - 50 = 4$. As another example, if $t = -4$, we get $(-12, 8)$, and we check: $3(-12) + 5(8) = -36 + 40 = 4$. As a final more extreme example, if $t = 123$, then $(8 + 5(123), -4 - 3(123) = (623, -373)$ is our new solution, and we check $3(623) + 5(-373) = 1869 - 1865 = 4$.

Knowing now how to solve linear equations in two unknowns, it is reasonable to ask about linear equations with three or more unknowns. There are numerous solution procedures one can employ, but we just mention one straight-forward procedure here.

Suppose we are given the equation $a_1x_1 + a_2x_2 + \cdots + a_kx_k = c$, with $k \geq 3$ and with $a_i \neq 0$ for all i.

(1) If $\gcd(a_1, a_2, \ldots, a_k) = d > 1$ and if d does not divide c, there are no solutions. Otherwise, just as we did above in the two unknowns case, divide through by d but label the equation $a_1x_1 + a_2x_2 + \cdots + a_kx_k = c$, where now $\gcd(a_1, a_2, \ldots, a_k) = 1$.

(2) The fact that $\gcd(a_1, a_2, \ldots, a_k) = 1$ means that there must exist (at least) two coefficients which are relatively prime. If necessary, rearrange the terms so that they are a_1 and a_2.

(3) Assign random values v_i to x_3 through x_k, so that we are now dealing with the following linear equation in two unknowns:

$$a_1x_1 + a_2x_2 = c - (a_3v_3 + \cdots + a_kv_k).$$

Now use our two unknowns procedure to solve this equation, obtaining as many solutions as you want (since that procedure already gives infinitely many solutions). Obviously, you can get more solutions by altering your random choices for the values of x_3 through x_k.

Example 9.2. Let's find just one solution of the equation

$$3x_1 + 5x_2 - 6x_3 + 2x_4 = 2.$$

Since $\gcd(3, 5) = 1$, we can proceed. Randomly, we set $x_3 = 1$ and $x_4 = 2$, so we are now trying to solve the equation

$$3x_1 + 5x_2 = 2 - ((-6)(1) + (2)(2)) = 4.$$

But we solved this equation in Example 9.1; our initial solution there was $x_1 = 8$ and $x_2 = -4$, so our solution to the current equation is $(8, -4, 1, 2)$.

We now move to the fascinating realm of non-linear Diophantine equations.

9.3   The Equation $x^2 + y^2 = z^2$

In this section we focus on what is likely the best known of all Diophantine equations: $x^2 + y^2 = z^2$. The reason is simple and is revealed in the following result, arguably the most famous theorem in all of mathematics. (For example, when the scarecrow gets his "brain" (actually a diploma) in the movie "The Wizard of Oz," he immediately states this theorem, although incorrectly, unfortunately.)

Theorem 9.1. (The Pythagorean Theorem) *Suppose that T is a triangle whose three sides are of lengths $\{x, y, z\}$. Then T is a right triangle with hypotenuse z if and only if $x^2 + y^2 = z^2$.*

There are many proofs of this theorem readily available, so here we concentrate instead on discovering what are the *integer* solutions of $x^2 + y^2 = z^2$. Such a set $\{x, y, z\}$ of integers is often called a *Pythagorean triple*. The simplest and best known Pythagorean triple is $\{3, 4, 5\}$ (since $9 + 16 = 25$); a couple of others are $\{5, 12, 13\}$ and $\{8, 15, 17\}$. Notice that in each of these three examples the hypotenuse z is odd and, of the two other sides, one is odd and the other is even. Will this pattern hold true in general?

We note that just as $\{3, 4, 5\}$ is a Pythagorean triple, so are $\{6, 8, 10\}$ (where $\gcd(6, 8) = 2$), $\{9, 12, 15\}$ (where $\gcd(9, 12) = 3$), and so on. In general, if $\{x, y, z\}$ is a Pythagorean triple with $\gcd(x, y) = 1$ and if d is a positive integer, then $\{dx, dy, dz\}$ is also a Pythagorean triple since $(dx)^2 + (dy)^2 = d^2(x^2 + y^2) = (dz)^2$. For this reason we will restrict our search for solutions to triples in which $\gcd(x, y) = 1$. Such triples are called *primitive Pythagorean triples*. The three examples given in the previous paragraph are primitive. A natural question is then, Are there infinitely many primitive Pythagorean triples? And also, Can we describe all of them? As we shall now see, the answer to both of these questions is "yes."

We assume from here on that the Pythagorean triple $\{x, y, z\}$ is primitive. This first says, of course, that x and y cannot both be even. Similarly, they cannot both be odd because we would have $x^2 \equiv 1 \pmod 4$ and $y^2 \equiv 1 \pmod 4$, so $z^2 \equiv 2 \pmod 4$, which is impossible when we recall from Chapter 7 that the only positive quadratic residue modulo 4 is 1. Hence we may and shall assume that y is even while x and z are both odd.

So with the goal of answering our two questions above, we examine the two numbers $(z + x)/2$ and $(z - x)/2$. We observe that

$$\frac{z+x}{2} + \frac{z-x}{2} = z \text{ and } \frac{z+x}{2} - \frac{z-x}{2} = x.$$

From these equations we see that $\gcd((z+x)/2, (z-x)/2)$ divides both z and x, which we are assuming to be relatively prime, and so $\gcd((z+x)/2, (z-x)/2) = 1$. But now

$$\frac{z+x}{2} \cdot \frac{z-x}{2} = \left(\frac{y}{2}\right)^2,$$

which tells us that $(z + x)/2$ and $(z - x)/2$ must both be perfect squares themselves. Hence we can write

$$(z + x)/2 = r^2 \text{ and } (z - x)/2 = s^2$$

for some positive integers r and s. It follows that r and s have these properties:

$$\gcd(r, s) = 1, \quad r > s, \quad x = r^2 - s^2, \quad y = 2rs \text{ and } z = r^2 + s^2.$$

Moreover, one of r and s must be even and the other odd, for otherwise both x and z would be even, which they are not.

We have shown that if we start with a primitive Pythagorean triple $\{x, y, z\}$, each number can be expressed in terms of some two positive integers r and s. Conversely, suppose that r and s are any two integers with $\gcd(r, s) = 1$, with $r > s > 0$, and with one even and one odd. If we now *define* $\{x, y, z\}$ by $x = r^2 - s^2$, $y = 2rs$, and $z = r^2 + s^2$, then we have that the integers x, y and z are positive, that $\gcd(x, y) = 1$, that x and z are odd, and that y is even. Finally, and importantly, we have

$$x^2 + y^2 = (r^2 - s^2)^2 + (2rs)^2 = (r^2 + s^2)^2 = z^2,$$

so $\{x, y, z\}$ is indeed a primitive Pythagorean triple. This then completes the proof of the following important theorem.

Theorem 9.2. *Every primitive Pythagorean triple (i.e., every primitive integer solution of the Diophantine equation $x^2 + y^2 = z^2$ with y even) is of the form*

$$x = r^2 - s^2, \quad y = 2rs \text{ and } z = r^2 + s^2,$$

where r and s are arbitrary integers, one even and one odd, satisfying $r > s > 0$ and $\gcd(r, s) = 1$.

Corollary 9.3. *There are infinitely many primitive Pythagorean triples.*

Proof. Since r and s can become arbitrarily large, the numbers x, y, and z can also become arbitrarily large. $\square$

Let's finish this section by listing a few relatively small primitive Pythagorean triples using Theorem 9.2.

r	s	$x = r^2 - s^2$	$y = 2rs$	$z = r^2 + s^2$
2	1	3	4	5
4	1	15	8	17
6	1	35	12	37
3	2	5	12	13
5	2	21	20	29
4	3	7	24	25
5	4	9	40	41

... and so on. Just checking the last one, for example, $9^2 + 40^2 = 81 + 1600 = 1681 = 41^2$. Even among this small number of triples, there are some interesting patterns which emerge and which we can explore further in the problem sections. See Problems 9.7, 9.8, 9.19 and 9.20.

9.4 The Equation $x^4 + y^4 = z^4$

Having just seen that the Diophantine equation $x^2 + y^2 = z^2$ has infinitely many primitive solutions (i.e., solutions in which $\gcd(x, y) = 1$), it would be natural to expect that the equations $x^3 + y^3 = z^3$, $x^4 + y^4 = z^4$, and so on would also have infinitely many solutions. As we shall now see in this section and the next, this turns out to be not true, and in fact very strongly not true in the sense that we go from infinitely many solutions to *no* (nontrivial) solutions for each power $n > 2$. This surprising fact was first conjectured in the 17-th century by Pierre de Fermat but was only proved in full generality about 350 years later in 1995 by Andrew Wiles. Wiles' proof uses methods which are *far* beyond elementary number theory, but we shall see here that with some considerable effort we are able to use our elementary techniques to conclude that the Diophantine equation $x^4 + y^4 = z^4$ has no integer solutions beyond the "trivial" ones; i.e., the ones in which x or y is 0. The exponent $n = 4$ turns out to be easier to deal with than $n = 3$ since with $n = 4$, we are able, as we shall soon see, to make use of Theorem 9.2.

To get the desired result about $x^4 + y^4 = z^4$, we will first closely examine the related equation $x^4 + y^4 = z^2$, after which we can apply what we learned about it to achieve our main goal. So here is our central result:

Theorem 9.4. *The only integral solutions of the Diophantine equation $x^4 + y^4 = z^2$ are the trivial solutions $\{x = 0, y, z = \pm y^2\}$ and $\{x, y = 0, z = \pm x^2\}$.*

Proof. Our proof illustrates the method of "proof by descent" or "Fermat's method of infinite descent," as it is sometimes called. The idea is the following: Assume that the equation $x^4 + y^4 = z^2$ has a non-trivial solution $\{x_0, y_0, z_0\}$ and then show that this assumption leads to another non-trivial solution $\{x_1, y_1, z_1\}$ with $z_1 < z_0$. Repeating this process will then lead to a solution with a smaller z_2, and so on. Since starting with z_0 cannot produce an infinite strictly decreasing sequence of positive integers, our assumption that the original solution existed must have been false, and the theorem follows. So get ready for the tricky argument needed to produce the first new "smaller" solution $\{x_1, y_1, z_1\}$. In the course of this proof it will be necessary to introduce eight new positive integers: $\{u, v\}$ and later $\{r, s\}$ when applying Theorem 9.2; $\{z_1, t\}$ and later $\{x_1, y_1\}$ when using the fact (already used once in the previous section) that if the product of two relatively prime integers is a perfect square, then each of the numbers must also be a perfect square.

So we now assume that we have a non-trivial, primitive solution $\{x_0, y_0, z_0\}$ of the equation $x^4 + y^4 = z^2$, and therefore $\{x_0^2, y_0^2, z_0\}$ is a primitive solution of the equation $x^2 + y^2 = z^2$. We may assume that y_0^2, and hence y_0, is even. Theorem 9.2 then tells us that there are integers u and v so that

$$x_0^2 = u^2 - v^2, \quad y_0^2 = 2uv, \quad z_0 = u^2 + v^2,$$

with $u > v > 0$, $\gcd(u, v) = 1$, and with u and v of opposite parity. In this case, however, if u were even, then v would be odd, and then we would have $x_0^2 \equiv 0 - 1 \equiv 3 \pmod 4$, which is a contradiction. Thus u must be odd and v must be even, so that

$$(\frac{y_0}{2})^2 = u \cdot \frac{v}{2} \text{ and } \gcd(u, \frac{v}{2}) = 1.$$

Hence u and $v/2$ must themselves be perfect squares, so for integers z_1 and t we have

$$u = z_1^2, \quad \frac{v}{2} = t^2 \text{ and } \gcd(z_1, t) = 1,$$

with $z_1 > 0$, $t > 0$, and z_1 odd.

Because $x_0^2 = u^2 - v^2$, we have $x_0^2 + v^2 = u^2$, so $x_0^2 + 4t^4 = z_1^4$. Hence we can again apply Theorem 9.2 to the Pythagorean triple $\{x_0, 2t^2, z_1^2\}$, which is primitive since $\gcd(z_1, 2t) = 1$. Thus there are integers r and s with the property that

$$x_0 = r^2 - s^2, \quad 2t^2 = 2rs, \quad z_1^2 = r^2 + s^2,$$

with $\gcd(r, s) = 1$ and $r > s > 0$. Since $rs = t^2$, we again have that both r and s are perfect squares, so we have $r = x_1^2$ and $s = y_1^2$ for some positive integers x_1 and y_1, with $\gcd(x_1, y_1) = 1$. But now we *finally* have

$$z_1^2 = x_1^4 + y_1^4;$$

that is, we have found a new primitive solution to the equation $x^4 + y^4 = z^2$, and we have that $z_1 < z_0$ because $z_0 = u^2 + v^2 = z_1^4 + v^2 > z_1^4 \geq z_1$.

As described at the beginning of this proof, since endlessly repeating this process of producing a new and "smaller" primitive solution from a current one is not possible, our original assumption of a primitive solution must have been false.

Finally, if $x^4 + y^4 = z^2$ has a solution in positive integers $\{a, b, c\}$ with $\gcd(a, b) = d > 1$, then $\{a/d, b/d, c/d^2\}$ would be a primitive solution since

$$d^4((a/d)^4 + (b/d)^4) = a^4 + b^4 = c^2 = d^4((c/d^2)^2).$$

(We note that since d^4 divides both a^4 and b^4, it must divide c^2 as well, so c/d^2 is indeed an integer.) However, we have shown that there are no primitive solutions, and hence the only solutions in integers to $x^4 + y^4 = z^2$ are the two trivial ones as stated in the theorem, and our proof is complete. $\square$

We now quickly get the result we originally sought:

Corollary 9.5. *The Diophantine equation $x^4 + y^4 = z^4$ has no non-trivial solutions.*

Proof. If $\{a, b, c\}$ were a non-trivial solution of $x^4 + y^4 = z^4$, then $\{a, b, c^2\}$ would be a non-trivial solution of $x^4 + y^4 = z^2$, which Theorem 9.4 tells us does not exist. Hence $\{a, b, c\}$ cannot be such a solution. $\square$

So what are we to make of the fact that the Diophantine equation $x^2 + y^2 = z^2$ has infinitely many integral solutions, but the equation $x^4 + y^4 = z^4$ has none? We'll now pin down the answer to this question.

9.5   The Equation $x^n + y^n = z^n$, $n > 2$

Having now closely examined the specific Diophantine equations $x^2 + y^2 = z^2$ and $x^4 + y^4 = z^4$, and having reached totally different conclusions about the existence of integral solutions of them, we now turn to the general case: the equation $x^n + y^n = z^n$. Here is what Pierre de Fermat concluded about its solutions.

Conjecture 9.6. *(Fermat's Conjecture, 1637) If $n > 2$, the Diophantine equation $x^n + y^n = z^n$ has no integral solutions except for the trivial solutions when one of x or y is 0.*

So Fermat stated that, even though $x^2 + y^2 = z^2$ has infinitely many integral solutions which we can explicitly describe (Theorem 9.2), *none* of the corresponding equations with exponents greater than 2 have *any* non-trivial solutions. (We proved in the previous section, with considerable effort, that this is true of the equation $x^4 + y^4 = z^4$.) Fermat claimed to have a proof of his conjecture, but he never wrote his proof down, if indeed he really did have one. Though various special cases of the theorem were settled in the ensuing years (for example, in 1770 Leonhard Euler gave a proof for the case $n = 3$), a general proof eluded mathematicians for centuries. Fermat's Conjecture was finally resolved in 1995, when Andrew Wiles at Princeton University gave an extremely complicated proof requiring several hundred pages of very advanced mathematics. There remains hope that someone might discover an *elementary* proof of Fermat's Conjecture (i.e., a proof using the kinds of arguments we used in the previous section to settle the matter for the case $x^4 + y^4 = z^4$), but this seems unlikely to occur because after more than 350 years no such elementary proof has been found.

We remark also that Fermat's Conjecture is often referred to as "Fermat's Last Theorem." It has been given this name since Fermat had a habit of writing down conjectures but then not supplying the needed proofs. One by one these other conjectures were proved to be true by other mathematicians as the years passed, but no proof of Conjecture 9.6 was found until 1995. Hence it was, for many, many years, his "last theorem."

9.6 Sums of Four Squares

In Section 9.3 we learned that the integer expression $x^2 + y^2$ can equal a perfect square z^2 for infinitely many triples $\{x, y, z\}$. In this section we ask a somewhat different question about sums of squares: *What is the smallest number k such that* every *positive integer can be written as a sum of k perfect squares?* (Note that we allow 0 to be one or more of the perfect squares in the list, since, after all, $0^2 = 0$.) Let's look at some examples.

Example 9.3. Since 3 can only be so written as $1+1+1$, k cannot be equal to 2. Now $4 = 0+0+4$, $5 = 0+1+4$ and $6 = 1+1+4$, *but* 7 must be $1+1+1+4$, so k cannot be 3 either. Going beyond 7 with $k = 4$: $8 = 0+0+4+4$, $9 = 0+0+0+9$, $10 = 0+0+1+9$, $11 = 0+1+1+9$, $12 = 1+1+1+9$, $13 = 0+0+4+9$, $14 = 0+1+4+9$, $15 = 1+1+4+9$, $16 = 0+0+0+16$, and so on. Things may be looking possible that $k = 4$. As just one larger example, $80 = 4+4+36+36$. We note also that such representations are not necessarily unique; for example, we also have $80 = 0+0+16+64$.

So we shall now embark on a series of arguments which will confirm that indeed $k = 4$; that is, every positive integer can be expressed as a sum of four perfect squares. We start with a lemma concerning the idea of *multiplicativity*. In Chapters 5 and 8, we called a numerical function f multiplicative if $f(nm) = f(n)f(m)$ for all integers n and m provided that $\gcd(n, m) = 1$, and *totally* multiplicative if $f(nm) = f(n)f(m)$ with no other conditions. What we shall see here is that the same concept can be applied to numerical *expressions* (such as a sum of four squares) as well as to numerical functions.

Leonard Euler (1707 - 1783), whom we just mentioned in Section 9.5, discovered the following complicated looking identity. In reality, the identity can be checked by simply multiplying out both sides and checking that the results are the same, but of course the computation is *very* laborious, so we omit the proof. What this result says, though, is that the product of two sums of four squares is itself a sum of four squares; that is, such expressions are multiplicative.

Lemma 9.7.

$$(x_1^2 + x_2^2 + x_3^2 + x_4^2)(y_1^2 + y_2^2 + y_3^2 + y_4^2) = (z_1^2 + z_2^2 + z_3^2 + z_4^2)$$

where

$$z_1 = x_1y_1 + x_2y_2 + x_3y_3 + x_4y_4, \quad z_2 = x_1y_2 - x_2y_1 + x_3y_4 - x_4y_3,$$

$z_3 = x_1y_3 - x_3y_1 + x_4y_2 - x_2y_4$ *and* $z_4 = x_1y_4 - x_4y_1 + x_2y_3 - x_3y_2$.

Example 9.4. Let's put one example into this lemma and see that it does work out. Since $5 = 0+0+1+4$ and $11 = 0+1+1+9$, we set $x_1 = x_2 = 0$, $x_3 = 1$, $x_4 = 2$ and $y_1 = 0$, $y_2 = y_3 = 1$, $y_4 = 3$. Now $z_1^2 + z_2^2 + z_3^2 + z_4^2$ becomes

$$(0 \cdot 0 + 0 \cdot 1 + 1 \cdot 1 + 2 \cdot 3)^2 + (0 \cdot 1 - 0 \cdot 0 + 1 \cdot 3 - 2 \cdot 1)^2$$

$$+ \ (0 \cdot 1 - 1 \cdot 0 + 2 \cdot 1 - 0 \cdot 3)^2 + (0 \cdot 3 - 2 \cdot 0 + 0 \cdot 1 - 1 \cdot 1)^2$$

$$= (0+0+1+6)^2 + (0-0+3-2)^2 + (0-0+2-0)^2 + (0-0+0-1)^2$$

$$= 49 + 1 + 4 + 1 = 55,$$

as expected! (Euler's coming up with this equality highlights his genius *and* persistence.)

Lemma 9.7 tells us that if the integers x and y can be written as sums of four squares, then the product xy can also be so written. Said another way, the expression of an integer as a sum of four squares is *totally multiplicative*. It follows that to prove that every positive integer can be expressed as a sum of four squares, *it suffices to prove that every prime number can be so expressed*. This fact will be of immense help in proving our main result, known as *Lagrange's Four Square Theorem*, named after Joseph-Louis Lagrange (1736 - 1813). Before being able to state and prove that theorem, we first require three preliminary results. The first and second ones, Lemmas 9.8 and 9.9, will help us with the difficult proof of Lemma 9.10. The following result covers one special case of the quantities in Lemma 9.7 which we will need to understand.

Lemma 9.8. *Suppose that the quantities* $x_1^2 + x_2^2 + x_3^2 + x_4^2$ *and* $y_1^2 + y_2^2 + y_3^2 + y_4^2$ *in Lemma 9.7 satisfy the following two conditions:*
    (1) *For some positive integer* m, $x_1^2 + x_2^2 + x_3^2 + x_4^2 \equiv 0 \pmod{m}$, *and*
    (2) *for each* $i = 1, \ldots, 4$, $y_i \equiv x_i \pmod{m}$.
Then for each $i = 1, \ldots, 4$, *the number* z_i *in Lemma 9.7 satisfies that* $z_i \equiv 0 \pmod{m}$.

Proof. By Condition 2 and then Condition 1 of the lemma,

$$z_1 = x_1y_1 + x_2y_2 + x_3y_3 + x_4y_4 \equiv x_1^2 + x_2^2 + x_3^2 + x_4^2 \equiv 0 \pmod{m}.$$

Now using only Condition 2, we have

$$z_2 = x_1y_2 - x_2y_1 + x_3y_4 - x_4y_3 \equiv x_1x_2 - x_2x_1 + x_3x_4 - x_4x_3 = 0 \pmod{m}.$$

The proofs for z_3 and z_4 are parallel to the one for z_2. $\square$

Now we come to our task of expressing a prime number p as a sum of four squares. This next result gets us nearer to (but not at) that goal.

Lemma 9.9. *If p is an odd prime, there is an integer m with $1 \leq m < (p-1)/2$ for which $mp = x_1^2 + x_2^2 + x_3^2 + x_4^2$ for some integers x_1, x_2, x_3, x_4.*

Proof. We first define the set S_1 as follows:

$$S_1 = \{0^2, 1^2, 2^2, \ldots, ((p-1)/2)^2\}.$$

Similarly, define the set S_2 by:

$$S_2 = \{-0^2 - 1, -1^2 - 1, -2^2 - 1, \cdots - ((p-1)/2)^2 - 1\}.$$

Note that for any two integers x and y, the congruence $x^2 \equiv y^2$ (mod p) implies that the prime p must divide $x - y$ or $x + y$. However the absolute values of both the sum and the difference of any two numbers x and y in S_1 are smaller than p and non-zero, so they are not divisible by p. We have then that their squares cannot be congruent modulo p. The exact same argument holds for the set S_2.

The two sets S_1 and S_2 contain a total of $p + 1$ distinct integers. We can now apply the "Box Principle" (or "Pigeon-Hole Principle"), which says that if you have n objects to be put into m boxes with $n > m$ (i.e., more objects than boxes), then at least one box must contain more than one object. In our application of the principle here, the boxes are the p congruence classes modulo p and the objects are the $p + 1$ integers in the sets S_1 and S_2 combined. We see then that some integer x^2 in S_1 and some $-y^2 - 1$ in

S_2, $x^2 \equiv -y^2 - 1 \pmod{p}$. We thus have $x^2 + y^2 + 1 \equiv 0 \pmod{p}$, with $0 \leq x \leq (p-1)/2$ and $0 \leq y \leq (p-1)/2$. This congruence means $x^2 + y^2 + 1 = mp$ for some integer m, and finally we have

$$1 \leq m = \frac{1}{p}(x^2 + y^2 + 1) \leq \frac{1}{p}(2(\frac{p-1}{2})^2 + 1) = \frac{p^2 - 2p + 3}{2p}$$

$$= \frac{p}{2} - 1 + \frac{3}{2p} \leq \frac{p}{2} - 1 + \frac{1}{2} = \frac{p-1}{2},$$

where in the last inequality we use the fact that $p \geq 3$. This completes the proof. $\square$

Notice that what we actually just proved can be stated more specifically as, If p is an odd prime, then there exist integers m, x, and y satisfying $1 \leq m \leq (p-1)/2$, $0 \leq x \leq (p-1)/2$ and $0 \leq y \leq (p-1)/2$ such that

$$mp = 0^2 + 1^2 + x^2 + y^2.$$

What we want, of course, is some similar statement in which $m = 1$.

Example 9.5. Letting $p = 7$, we know that the only representation of 7 as a sum of four squares is $7 = 1^2 + 1^2 + 1^2 + 2^2$, so in this case m cannot be 1. However, $m = 2$ gives us $14 = 0^2 + 1^2 + 2^2 + 3^2$. The case $m = 3$ gives us $21 = 0^2 + 1^2 + 2^2 + 4^2$, but 4 does not satisfy the condition that y is less than or equal to 3. Hence for $p = 7$, $m = 2$ is the only case which "works." In particular, the desired case of $m = 1$ does not work.

We are now able to find a way to force m to be 1. By the multiplicativity of four squares representations proved in Lemma 9.7, we will thus achieve our desired result.

Lemma 9.10. *If m is the least integer satisfying* Lemma 9.9, *then* $m = 1$.

Proof. We first show that if m is even, then it cannot be minimal. Assuming that m is even, then so is $mp = x_1^2 + x_2^2 + x_3^2 + x_4^2$, so that either zero, two, or four of the x_i are even. If exactly two of the x_i are even, we can renumber the x_i so that x_1 and x_2 are the two even ones. Then in any case (i.e., zero, two or four even x_i) $x_1 \pm x_2$ and $x_3 \pm x^4$ are even. Hence we have

$$(\frac{x_1 + x_2}{2})^2 + (\frac{x_1 - x_2}{2})^2 + (\frac{x_3 + x_4}{2})^2 + (\frac{x_3 - x_4}{2})^2 = \frac{m}{2}p.$$

Since $m/2$ also gives us a four squares representation, m was not minimal, and so a minimal m cannot be even.

We now assume that $m > 1$ and hope to reach a contradiction, which will force $m = 1$. Because we know that 2, 3, and 5 can be written as a sum of four squares, we assume that $p \geq 7$. Since m is odd, we have $3 \leq m \leq (p-1)/2$. For $i = 1, \ldots, 4$, define the numbers y_i by $y_i \equiv x_i \pmod{m}$ with $-\frac{m-1}{2} \leq y_i \leq \frac{m-1}{2}$ (so we're replacing, if necessary, x_i with a number y_i which is congruent to x_i modulo m but has a "small" absolute value). Since $x_1^2 + \cdots + x_4^2 = mp$, we have $y_1^2 + \cdots + y_4^2 \equiv x_1^2 + \cdots + x_4^2 \equiv 0 \pmod{m}$. Thus we can write

$$y_1^2 + \cdots + y_4^2 = mn \text{ with } 0 \leq n \leq \frac{4}{m}(\frac{m-1}{2})^2$$

$$= \frac{m^2 - 2m + 1}{m} = m - 2 + \frac{1}{m} < m.$$

We need now to eliminate the possibility that $n = 0$. If n were 0, we would have $y_1 = \cdots = y_4 = 0$ and hence by definition of the y_i, m divides each x_i, so m^2 divides each x_i^2. This would imply then that $mp = x_1^2 + \cdots + x_4^2 \equiv 0 \pmod{m^2}$, i.e., we would have that m^2 divides mp, but this is not possible since $3 \leq m \leq (p-1)/2 < p$ and so m^2 cannot divide the prime p. Hence we have shown that $n > 0$.

We are almost there. From Lemma 9.7 we have that

$$m^2 np = (mp)(mn) = (x_1^2 + \cdots + x_4^2)(y_1^2 + \cdots + y_4^2) = z_1^2 + \cdots + z_4^2,$$

for some integers z_i. Lemma 9.8 now tells us that $z_i \equiv 0 \pmod{m}$ for $i = 1, \ldots, 4$. Dividing by m^2, we obtain

$$np = (\frac{z_1}{m})^2 + \cdots + (\frac{z_4}{m})^2$$

where $0 < n < m$. This proves that m is not minimal if $m > 1$. Hence $m = 1$ and the proof is complete. $\square$

Thus we finally arrive at a very nice result in number theory.

Theorem 9.11. (*Lagrange's Four Square Theorem*) *Every positive integer can be written as a sum of four squares, and no number smaller than four has this property.*

Proof. By Lemma 9.7 it suffices to show that every prime number p can be so expressed. By Lemma 9.10 this holds for all odd primes. Since $1 = 0^2 + 0^2 + 0^2 + 1^2$ and $2 = 0^2 + 0^2 + 1^2 + 1^2$, we are done. $\square$

To complete this section, we introduce a function, attributed to Carl Gustav Jacob Jacobi (1804 - 1851), which determines the number of ways that a given positive integer can be written as a sum of four squares, provided that we count rearrangements as being different and allow negative as well as positive entries. We state the theorem here without proof.

Theorem 9.12. (*Jacobi's Four Square Theorem*) *Let $\tau_4(n)$ denote the number of ways of expressing n as a sum of four squares. Then*

$$\tau_4(n) = \begin{cases} 8 \sum_{m|n} m & \text{if } n \text{ is odd,} \\ 24 \sum_{\substack{m|n \\ m \text{ odd}}} m & \text{if } n \text{ is even.} \end{cases}$$

Example 9.6. Let's examine this formula for three cases: $n = 1$, 4, and 7.
(a) For $n = 1$, the formula predicts $8 \sum_{m|1} m = 8(1) = 8$. Here are 3 of those 8:

$$1 = 1^2 + 0^2 + 0^2 + 0^2 = 0^2 + 1^2 + 0^2 + 0^2 = (-1)^2 + 0^2 + 0^2 + 0^2.$$

The other 5 should be easy to see. The answer is 8 since there are 4 places to put the 1^2 and for each of those the choice between 1^2 and $(-1)^2$.

(b) For $n = 4$, the formula predicts $24 \sum_{m|4, m \text{ odd}} m = 24(1) = 24$. The number 4 has two distinct "types" of four square representations: $1^2 + 1^2 + 1^2 + 1^2$ and $2^2 + 0^2 + 0^2 + 0^2$. For the former type we just need to count how many ways there are to insert $(-1)^2$. Well, there's 1 way to insert none, 4 ways to insert one, 6 ways to insert two, 4 ways to insert three, and 1 way to insert four. That gives us 16 representations of this type. For the other type, the 2^2 can be put in 4 places, and likewise for $(-2)^2$, so we get 8 more representations for a total of 24, as predicted.

(c) For $n = 7$, the formula predicts $8 \sum_{m|7} m = 8(1+7) = 64$. The only type of representation for 7 is, as we have seen, $2^2 + 1^2 + 1^2 + 1^2$. We have 4 ways to place the 2^2, and likewise for the $(-2)^2$, so that's 8 so far. For *each* of these we can insert $(-1)^2$ into the remaining 3 places as follows: 1 way to insert none, 3 ways to insert one, 3 ways to insert two, and 1 way to insert three, so that's a total of 8. Hence we get $8 \cdot 8 = 64$ representations, as predicted.

9.7   Waring's Problem

Once again, having studied an idea involving perfect squares, it is natural to ask what happens with cubes, fourth powers, and so on. In 1770 Edward Waring (1736 - 1798) asked whether the four square theorem can be extended to higher powers. He conjectured that nine cubes would suffice (i.e. every positive integer can be written as a sum of nine cubes), and that 19 fourth powers would suffice. David Hilbert (1862 - 1943) proved that for each positive integer k, there is a number $g(k)$ such that every positive integer can be written as a sum of $g(k)$ k-th powers, and no number smaller than $g(k)$ will suffice. We know that $g(1) = 1$ since any positive integer can be written as itself, and we proved in the previous section that $g(2) = 4$. It is now also known that $g(3) = 9$, $g(4) = 19$, $g(5) = 37$, and $g(6) = 73$, so Waring was indeed correct in his conjectures.

Is there a general formula for $g(k)$? The following formula has been conjectured but not as yet proved.

Conjecture 9.13.

$$g(k) = 2^k + \left[(\frac{3}{2})^k\right] - 2,$$

for all positive integers k, where [x] denotes the greatest integer function.

Example 9.7. Let's substitute a few values into the conjectured formula for $g(k)$ and see if they match up with the values which have been proven:

For $k = 1$, $2^1 + [(3/2)^1] - 2 = 2 + 1 - 2 = 1$,

for $k = 2$, $2^2 + [(3/2)^2] - 2 = 4 + [9/4] - 2 = 4 + 2 - 2 = 4$,

for $k = 3$, $2^3 + [(3/2)^3] - 2 = 8 + [27/8] - 2 = 8 + 3 - 2 = 9$,

for $k = 4$, $2^4 + [(3/2)^4] - 2 = 16 + [81/16] - 2 = 16 + 5 - 2 = 19$,

for $k = 5$, $2^5 + [(3/2)^5] - 2 = 32 + [243/32] - 2 = 32 + 7 - 2 = 37$,

and

for $k = 6$, $2^6 + [(3/2)^6] - 2 = 64 + [729/64] - 2 = 64 + 11 - 2 = 73$.

This is remarkable *evidence* that the conjecture is true, but it is not, of course, *proof* that it is true for all positive integers k.

9.8 Summary

Diophantine equations, i.e., equations whose coefficients and possible solutions must be integers, possess properties and suggest problems which are unique in mathematics. In this chapter we learned how to solve (if possible) general linear Diophantine equations, and then we focused on two problems involving non-linear equations, specifically:

(1) Does the equation $x^n + y^n = z^n$ have any (non-trivial) solutions, and if so can we describe them? The surprising answer turns out to be:

(a) If $n = 2$, there are infinitely many non-trivial solutions, and we can fully describe them, but

(b) if $n > 2$, there are *no* non-trivial solutions. We proved that this is the case for the equation $x^4 + y^4 = z^4$.

(2) Can every positive integer be written as a sum of some smallest fixed number of k-th powers of integers? The answer is "yes!",

where that smallest fixed number depends on k (for example, for $k = 2$, we proved that that number is 4).

There are of course many other such challenging problems concerning Diophantine equations, but this is as far as we choose to go in this text. For just one example, an unsolved problem involves "Kraus' equation": Solve $x^3 + y^3 = z^p$ for all primes $p \geq 3$. As we learned, the case $p = 3$ has been solved (no solutions), but what about other values of p? And the list of possible questions about Diophantine equations, answered and unanswered, goes on and on.

9.9   Solved Problems

Linear Equations

9.1. Find all solutions of the linear equation $10x - 6y = 17$.

Solution:
Since $\gcd(10, 6) = 2$ and 2 does not divide 17, there are no solutions.

9.2. Find all solutions of the linear equation $10x - 7y = 17$.

Solution:
Since $\gcd(10, 7) = 1$, there are solutions. By inspection, $x = 1$ and $y = -1$ is a solution. By Step (4) of our procedure in Section 9.2, the general solution is $(1 + (-7)t, -1 - 10t) = (1 - 7t, -1 - 10t)$. For example, if $t = 1$, we have the solution $(-6, -11)$, and we can check that $10(-6) - 7(-11) = -60 + 77 = 17$.

9.3. Find all solutions of the linear equation $10x - 6y = 18$.

Solution:
Since $\gcd(10, 6) = 2$ and 2 divides 18, we can simplify to the equivalent equation $5x - 3y = 9$, which has solutions since $\gcd(5, 3) = 1$. Since a solution is not evident by inspection, we first solve the equation $5x - 3y = 1$, which by inspection has a solution $(2, 3)$. Multiplying through by 9, we get $5(9 \cdot 2) - 3(9 \cdot 3) = 9$, i.e., we have

found the solution $(18, 27)$. We can check: $5(18)-3(27) = 90-81 = 9$. Hence the general solution is $(18 - 3t, 27 - 5t)$. Checking one of these solutions, if, say, $t = 2$, we get the solution $(12, 17)$, and we check that $5(12) - 3(17) = 60 - 51 = 9$.

9.4. Find one solution of the linear equation $2x_1+4x_2-6x_3+5x_4 = 10$.

Solution:
Since the coefficients 2 and 5 of x_1 and x_4 are relatively prime, we assign random values to x_2 and x_3, say $x_2 = x_3 = 1$ and now solve the equation $2x_1 + 5x_2 = 10 - 4(1) + (6)1 = 12$. A solution here is $x_1 = 1$ and $x_4 = 2$, so our solution (one of many!) is $(1, 1, 1, 2)$. We check: $2(1) + 4(1) - 6(1) + 5(2) = 10$.

9.5. Suppose that $ax + by = c$ with $\gcd(a, b) = 1$, and suppose there are two solutions, (x_0, y_0) and (x_1, y_1) with $x_1 = 1 + x_0$. Prove that $b = \pm 1$.

Solution:
We have

$$ax_0 + by_0 = ax_1 + by_0 = a(x_0 + 1) + by_1 = ax_0 + a + by_1,$$

and so $b(y_0 - y_1) = a$, which shows that b divides a. Hence $b = \gcd(a, b) = 1$ if $b > 0$ and $-b = \gcd(a, b) = 1$ if $b < 0$, i.e., $b = \pm 1$.

The Equation $x^2 + y^2 = z^2$

9.6. Continuing with the table at the end of Section 9.3, use Theorem 9.2 to write down the two "smallest" primitive Pythagorean triples for which $s = 5$.

Solution:
$r = 6$, $s = 5$ gives $x = 36 - 25 = 11$, $y = 2 \cdot 6 \cdot 5 = 60$ and $z = 36 + 25 = 61$,
$r = 8$, $s = 5$ gives $x = 64 - 25 = 39$, $y = 2 \cdot 8 \cdot 5 = 80$ and $z = 64 + 25 = 89$.

9.7. Notice in the table of Section 9.3, one of x or y is divisible by 3. Prove that this is always the case in primitive Pythagorean triples.

Solution:
Suppose that neither x nor y is divisible by 3. Recall from Chapter 7 that we must then have $x^2 \equiv y^2 \equiv 1 \pmod 3$, which forces $z^2 \equiv 2$, which is a contradiction. Hence one of x or y is divisible by 3.

9.8. Prove that there are infinitely many primitive Pythagorean triples in which the hypotenuse z is 2 more than the odd side x.

Solution:
Notice that the first three entries in the Section 9.3 table have this property. For any integers r and s satisfying the hypotheses of Theorem 9.2, $z = x + 2$ implies that $r^2 + s^2 = r^2 - s^2 + 2$, so $s^2 = 1$, so $s = 1$. The next such triple not in the table would be generated by $r = 8$, $s = 1$, and so would be $(63, 16, 65)$.

Sums of Squares

9.9. Write 29, 56, 110, and 127 as sums of four squares.

Solution:
$$29 = 5^2 + 2^2 + 0^2 + 0^2$$
$$56 = 6^2 + 4^2 + 2^2 + 0^2$$
$$110 = 10^2 + 3^2 + 1^2 + 0^2$$
$$127 = 10^2 + 5^2 + 1^2 + 1^2$$

9.10. Counting distinct representations as in Jacobi's Four Square Theorem (Theorem 9.12), determine the number of different ways of expressing 29 and 30 as sums of four squares.

Solution:
Since 29 is odd and prime, we have that the total number of such representation is $8(1 + 29) = 8(30) = 240$.

Since 30 is even and its odd divisors are $\{1, 3, 5, 15\}$, we have that the total number of representations is $24(1+3+5+15) = 576$.

Waring's Problem

9.11. Write 157 as a sum of perfect cubes.

Solution:

$$5^3 + 3^3 + 1^{3\cdot} + 1^3 + 1^3 + 1^3 + 1^3$$

9.12. According to Conjecture 9.13, what is the value of $g(7)$?

Solution:

$$g(7) = 2^7 + [(3/2)^7] - 2 = 128 + [2187/128] - 2 = 128 + 17 - 2 = 143.$$

9.10 Supplementary Problems

9.13. Find all solutions of the linear equation $8x - 6y = 11$.

9.14. Find all solutions of the linear equation $8x - 5y = 11$.

9.15. Find all solutions of the linear equation $8x - 6y = 12$.

9.16. Find one solution of the linear equation $2x_1 - 4x_2 + 3x_3 + 6x_4 = 10$.

9.17. Prove that the equation $ax + by = a + c$ is solvable in integers if and only if the equation $ax + by = c$ is solvable in integers.

The Equation $x^2 + y^2 = z^2$

9.18. Continuing with the table at the end of Section 9.3, use Theorem 9.2 to write down the two "smallest" primitive Pythagorean triples for which $s = 6$.

9.19. Notice in the table of Section 9.3 that one of x, y or z is divisible by 5. Prove that this is always the case in primitive Pythagorean triples. (Suggestion: See Problem 9.7 and its solution. If z is divisible by 5 we are done, so assume z is not divisible by 5 and can go from there.)

9.20. (a) Prove that every odd number $n \geq 3$ appears as the side x in a primitive Pythagorean triple. (Suggestion: Look at the Section 9.3 table.)
(b) Prove that every number m which is divisible by 4 appears as the side y in a primitive Pythagorean triple.
(c) Give an example of a number $k \geq 3$ which cannot appear in a primitive Pythagorean triple and justify your answer.

The Equation $x^n + y^n = z^n$

9.21. Prove that the Diophantine equation $x^{100} + y^{100} = z^{100}$ has no non-trivial solutions.

Sums of Squares and Higher Powers

9.22. Write each of the eight numbers 40 through 47 as a sum of squares, using as few non-zero summands as possible.

9.23. Notice that each of the examples in the previous problem required only three or fewer non-zero squares except for 47. Prove that no integer of the form $8k + 7$ can be expressed as a sum of three squares. (Suggestion: Show that for any positive integer a, $a^2 \equiv 0, 1$ or 4 (mod 8). Now, what numbers can any three of $\{0, 1, 4\}$ add up to? For example, $0 + 1 + 4 = 5$, $1 + 1 + 1 = 3$, etc.)

9.24. Counting distinct representations as in Jacobi's Four Square Theorem (Theorem 9.12), determine the number of different ways of expressing 27 and 34 as sums of four squares.

Waring's Problem

9.25. Write 188 as a sum of cubes.

9.26. According to Conjecture 9.13, what is the value of $g(8)$?

Answers to Selected Supplementary Problems

9.13. None.

9.14. $(2 - 5t, 1 - 8t)$

9.15. $(6 - 3t, 6 - 4t)$

9.16. $(2, 1, 3, 1)$

9.18. For $r = 7$ and $s = 6$, we get $x = 16$, $y = 84$, $z = 85$.
For $r = 11$ and $s = 6$, we get $x = 85$, $y = 132$, $z = 157$.

9.22.

$$40 = 6^2 + 2^2 + 0^2 + 0^2$$
$$41 = 5^2 + 4^2 + 0^2 + 0^2$$
$$42 = 5^2 + 4^2 + 1^2 + 0^2$$
$$43 = 5^2 + 3^2 + 3^2 + 0^2$$
$$44 = 6^2 + 2^2 + 2^2 + 0^2$$
$$45 = 6^2 + 3^2 + 0^2 + 0^2$$
$$46 = 6^2 + 3^2 + 1^2 + 0^2$$
$$47 = 6^2 + 3^2 + 1^2 + 1^2$$

9.24. 27 has 320 representations; 34 has 432.

9.25. $5^3 + 3^3 + 3^3 + 2^3 + 1^3$

9.26. $g(8) = 279$.

Chapter 10

Finite Fields

10.1   Introduction

In Chapter 3 we introduced the sets $\mathbb{Z}_n$, whose elements are the n numbers $\{0, 1, 2, \ldots, n-1\}$ and in which we can perform the algebraic operations of addition, subtraction and multiplication, provided that these operations are done modulo n. This way the results of all operations in $\mathbb{Z}_n$ are themselves in $\mathbb{Z}_n$. We say then that the sets $\mathbb{Z}_n$ are "finite sets with algebraic structure." Moreover, we also learned in that chapter that we can sometimes perform the operation of division in $\mathbb{Z}_n$; specifically, if an element a in $\mathbb{Z}_n$ possesses a multiplicative inverse a^{-1} in $\mathbb{Z}_n$ (i.e., $a \cdot a^{-1} = 1$), then b "divided by" a is just the product $b \cdot a^{-1}$. So under what conditions does $a \in \mathbb{Z}_n$ possess a multiplicative inverse? The answer was given in Lemma 3.3: $a \in \mathbb{Z}_n$ has a multiplicative inverse in $\mathbb{Z}_n$ if and only if $\gcd(a, n) = 1$. It follows then that we can perform division by every non-zero element of $\mathbb{Z}_n$ if and only if $n = p$, a prime number.

Mathematicians refer to a set F which has algebraic structure (i.e., addition, subtraction, and multiplication) and in which every non-zero element has a multiplicative inverse in F as a *field*. Thus $\mathbb{Z}_p$ is a field for every prime p. Since $\mathbb{Z}_p$ is also finite, it is, quite naturally, called a *finite field*. Examples of *infinite* fields with which you are familiar are the set $\mathbb{R}$ of real numbers and the set $\mathbb{Q}$ of

DOI: 10.1201/9781003193111-10

rational numbers (i.e., fractions) since for every non-zero real or rational number x, $1/x$ is its multiplicative inverse.

In this chapter, then, we will concentrate on finite fields, and we already have an infinite number of examples (i.e., $\mathbb{Z}_p$) since there are infinitely many prime numbers p. We note that if n is composite, $\mathbb{Z}_n$ is definitely *not* a finite field since if $0 < a < n$ is in $\mathbb{Z}_n$ with $\gcd(a, n) \neq 1$, Lemma 3.3 tells us that a has no multiplicative inverse in $\mathbb{Z}_n$. A central question becomes, Do there exist other finite fields besides the collection $\{\mathbb{Z}_p | p \text{ prime}\}$. The answer is yes, as we shall now learn.

10.2   The Finite Fields $\mathbb{F}_{p^n}$

To emphasize that we have a field, from here on we shall denote the finite field with p elements by $\mathbb{F}_p$ rather than $\mathbb{Z}_p$. This is simply a notational change; $\mathbb{F}_p$ and $\mathbb{Z}_p$ both denote the exact same set with the exact same algebraic structure.

By a *prime power* we mean a positive integer of the form p^e where p is a prime and $e \geq 1$ is a positive integer. For example, 2^4, 5^{10}, and 97^2 are prime powers, but numbers like 6, 10, 12, 24, 96 are not prime powers. We note that prime numbers are also prime powers since the exponent e can be 1. Below we shall often use q to represent a prime power; that is, $q = p^e$ for some prime p and some positive integer exponent e.

Here then is the answer to our question above about the existence of finite fields besides the collection $\{\mathbb{F}_p | p \text{ prime}\}$.

Theorem 10.1. (a) *If p is a prime and $e \geq 1$ is a positive integer, there is a finite field $\mathbb{F}_{p^e}$ of order p^e, i.e., which contains exactly p^e distinct elements.*
(b) *Moreover if F is a finite field, then F must contain exactly p^e distinct elements for some prime p and some positive integer $e \geq 1$.*

Thus, for example, there are finite fields of orders 2, 2^2, 2^{10}, 5^8, and 97^6 but there are no finite fields of orders 6, 10, 12, 24, or 96.

We shall prove Part (b) of Theorem 10.1 in the following section, and in Section 10.4 we shall prove Part (a).

There are several notations to represent finite fields. A common notation for the finite field containing q elements is $GF(q)$ where G stands for Galois and F stands for field. This name is used in honor of Evariste Galois (1811 – 1832), who in 1830 was the first person to closely study general finite fields (i.e., fields with a prime power but not a prime number of elements). We will however use the slightly simpler notation $\mathbb{F}_q$ to represent the finite field with exactly q elements. By Theorem 10.1, q must be a prime power.

Although (as we pointed out above) the finite field $\mathbb{F}_p$ is identical to the set $\mathbb{Z}_p$ for all primes p, this is *not* true for the field $\mathbb{F}_q$ and the set $\mathbb{Z}_q$ when $q = p^e$ with $e > 1$. This latter set cannot be a field since, by Lemma 3.3, $\mathbb{Z}_q$ contains non-zero elements which do not possess a multiplicative inverses. For example, the elements 2, 3, and 4 in $\mathbb{Z}_6$ have no multiplicative inverses, and the same is true of 2, 4, and 6 in $\mathbb{Z}_8$. In Problem 10.1 you are asked to look closely at the cases of the field $\mathbb{F}_4$ and the non-field $\mathbb{Z}_4$, and likewise in Problem 10.12 for $\mathbb{F}_9$ and $\mathbb{Z}_9$.

10.3   The Order of a Finite Field

From our previous work we know exactly the structure of the finite field $\mathbb{F}_p$ where p is prime. Specifically, this field has p elements denoted $\{0, 1, \ldots, p - 1\}$ and its operations of addition and multiplication are carried out modulo p. However, if F is a finite field but is not one of these, we know little about it except that its *order* (i.e, its number of elements) is finite. Theorem 10.1, which we stated without proof (as yet), says that this order is a prime power, but we need to prove this fact. Moreover, we know little about the algebraic structure of F; that is, how do we perform addition and multiplication? In this section we will concentrate on the former question about order, and then in Section 10.4 we'll turn to the question about structure.

In order to explore these matters, we need to be reminded of (or learn) some basic ideas from linear and abstract algebra. Our intent here is not to provide every detail but only to provide the necessary basics.

(1) If F is any field and K is a subset of F, then K is a *subfield* of F if K is itself a field under the operations of F. So, for example, the set $\mathbb{Q}$ of rational numbers is a subfield of the set $\mathbb{R}$ of real numbers since the sum, difference, product, and quotient of two rational numbers is a rational number. We note in passing that the set $\mathbb{Z}$ of integers is *not* a subfield of $\mathbb{R}$ since $\mathbb{Z}$ is not itself a field (only 1 and -1 have multiplicative inverses in $\mathbb{Z}$).

(2) If K is any field, a *vector space* V over K is a set which has an addition operation (in the language of abstract algebra, it is an "additive Abelian group"), and it also has a "scalar multiplication" by elements of K; that is, if v is in V and λ is in K, then the product λv must also be in V. If there is a set $B = \{b_1, \ldots, b_n\}$ of elements of V such that every $v \in V$ has a *unique* representation $v = \lambda_1 b_1 + \cdots + \lambda_n b_n$ (where each λ_i is in K), then we say that B is a *basis* for V over K and we say that the *dimension* of V over K is n.

(3) The *characteristic* of a field F is the smallest number k such that

$$\underbrace{1 + 1 + \cdots + 1}_{k \text{ terms}} = 0.$$

If no such k exists, we say that the characteristic is 0; for example, the set of real numbers $\mathbb{R}$ is an infinite field of characteristic 0. Also note that k, if it exists, must be prime, for if $k = a \cdot b$ with $a, b > 1$, then

$$0 = \underbrace{1 + 1 + \cdots + 1}_{k \text{ terms}} = (\underbrace{1 + 1 + \cdots + 1}_{a \text{ terms}})(\underbrace{1 + 1 + \cdots + 1}_{b \text{ terms}}),$$

so one of these factors must be 0, contradicting the minimality of k.

We can now use the concepts of subfield and vector space to pin down the possible order of a finite field, as stated in Theorem 10.1. Below, the notation $|F|$ means the order of F; i.e., the number of elements in F.

Lemma 10.2. *Assume F is a finite field with a subfield K containing s elements. Then F is a vector space over K and $|F| = s^n$, where n is the dimension of F over K.*

Proof. Since F is a field, it is already equipped with an addition operation, and moreover if v is in F and λ is in K, then the product λv is defined and lies in F. As part of the definition of a vector space, we must require K itself to be a field because if λ and δ are in K, we must have $\lambda(\delta v) = (\lambda\delta)v$ and $(\lambda + \delta)v = \lambda v + \delta v$. Hence F is a vector space over K. Since F is finite, we can choose a finite basis $B = \{b_1, \ldots, b_n\}$ for F over K. Every element v of F can thus be written uniquely in the form $v = \lambda_1 b_1 + \cdots + \lambda_n b_n$, where each λ_i is in K and the sequence $\lambda_1, \ldots, \lambda_n$ is uniquely determined by v. Since there are $|K| = s$ choices for each λ_i, there are $|K|^n = s^n$ distinct sequences of coefficients. This completes the proof. $\square$

We would like now to show that if F is an arbitrary finite field, it will contain $\mathbb{F}_p$ (for some prime p) as a subfield. This is where the concept of the characteristic of a field comes in.

Lemma 10.3. *If F is a finite field, then, for some prime p, the characteristic of F is p. Hence F contains $\mathbb{F}_p$ as a subfield.*

Proof. Since F is finite, it cannot be of characteristic 0 (otherwise the infinite set $\{1, 2, 3, \ldots\}$ would be a subset of F). Hence, by our paragraph (3) above, F must be of characteristic p for some prime p. This tells us that the set $\{0, 1, 2, \ldots, p - 1\}$ is a subset of F, and it also tells us that the addition and multiplication within this set are being done modulo p. Hence the subset $\{0, 1, 2, \ldots, p - 1\}$ with these operations is in fact $\mathbb{F}_p$, and we are done. $\square$

Proof of Part (b) of Theorem 10.1. Let F be an arbitrary finite field. By Lemma 10.3, F contains $\mathbb{F}_p$ (for some prime p) as a subfield. Then by Lemma 10.2, $|F| = p^e$, where e is the dimension of F over $\mathbb{F}_p$. This completes the proof. $\square$

What we do not yet know, as claimed in Theorem 10.1, Part (a), is that for *every* prime p and *every* integer exponent $e > 1$, there is a finite field F with $|F| = p^e$. We shall address this matter in the next section, which will also show us a way to construct the elements and operations of F.

10.4   Constructing Finite Fields

We now turn to the following question: Given a prime number p and a positive integer e, how can we construct both the elements and the arithmetic operations of a finite field $\mathbb{F}_{p^e}$? Of course we've long since known how to do this when $e = 1$, but we have to this point not considered how to do it when $e > 1$. It turns out that a convenient way to do this construction is to utilize polynomials in a single unknown whose coefficients are taken from the prime finite field $\mathbb{F}_p$. We shall denote by $\mathbb{F}_p[\theta]$ the set of all polynomials in the single unknown θ with coefficients in $\mathbb{F}_p$, where p is any prime. This set has algebraic structure by using standard addition and multiplication of polynomials. Here are two important definitions:

A polynomial $f(\theta)$ is called *monic* if its leading term (i.e., non-zero term of highest degree) has a coefficient of 1. So, for example $\theta^2 + 4\theta + 3$ is monic but $2\theta + 4$ is not. Important idea: *Monic polynomials in $\mathbb{F}_p[\theta]$ are the analogue of positive numbers in the integers $\mathbb{Z}$.*

A polynomial $f(\theta)$ is called *irreducible* if it cannot be factored into two polynomials of positive degree. For example, in $\mathbb{F}_2[\theta]$, $\theta^2 + \theta + 1$ is irreducible, but $\theta^2 + 1 = (\theta + 1)(\theta + 1)$ is not. We note that the irreducibility of a polynomial depends on the field of coefficients. For example, the polynomial $x^2 - 2$ is irreducible over the field of

$\mathbb{Q}$ of rational numbers, but it is *reducible* over the field $\mathbb{R}$ of real numbers since there $x^2 - 2 = (x - \sqrt{2})(x + \sqrt{2})$. Important idea: *Irreducible polynomials in $\mathbb{F}_p[\theta]$ are the analogue of prime numbers in the integers $\mathbb{Z}$.*

Given these two definitions, we can now describe a procedure for constructing a finite field F whose order is p^e with $e > 1$.

1. Select a monic irreducible polynomial $P(\theta)$ of degree e in $\mathbb{F}_p[\theta]$. We shall show in Section 10.7 that there do exist monic irreducible polynomials of all degrees in $\mathbb{F}_p[\theta]$; in fact we shall be able to count exactly how many there are of each degree.

2. Let F be the set of all polynomials (not just monic ones) of degree $e - 1$ or less in $\mathbb{F}_p[\theta]$. Since there are e coefficients to fill in with p choices in each one, there are a total of p^e elements (which are polynomials in θ) in F, so F now has the correct number of elements.

3. The algebraic structure of F is now the following: Addition is the standard polynomial term-by-term operation, so F is definitely closed under this operation. Multiplication is done by standard polynomial multiplication, *but* is followed by reducing the answer modulo our monic irreducible polynomial $P(\theta)$ defined above. That is, we divide our answer by $P(\theta)$ and take the remainder, whose degree must be less than or equal to $e - 1$ since the degree of $P(\theta)$ is e.

Thus F is of order p^e and is equipped with addition and multiplication, but why is F a *field*? The abstract algebra we need is beyond what we wish to do, but here is the analogy with the $\mathbb{Z}_n$ case: In $\mathbb{Z}_n$ we get a field if and only if the modulus is prime, but the analogue of "prime" in $\mathbb{Z}$ is "irreducible" in $\mathbb{F}_p[\theta]$, so our procedure above does indeed guarantee that F is a field, which we now denote by $\mathbb{F}_{p^e}$. (For those readers who have studied abstract algebra, the theorem we are applying here is, If R is a commutative ring with unity and if I is a maximal ideal in R, then the factor ring R/I is a field.)

We finally have then:

Proof of Theorem 10.1, Part (a). We have just shown, given any prime number p and any positive integer e, how to construct a finite field, denoted $\mathbb{F}_{p^e}$, whose order is p^e. $\square$

Here now are two examples of our construction:

Example 10.1. First the field $F_4 = F_{2^2}$:
Since $e = 2$, we choose $P(\theta) = \theta^2 + \theta + 1$ as our monic irreducible polynomial from $F_2[\theta]$. (In fact this is the only irreducible quadratic polynomial in $F_2[\theta]$, see Section 10.7.) Hence the four elements of F_4 are $\{0, 1, \theta, \theta + 1\}$. The addition table is:

+	0	1	θ	$\theta+1$
0	0	1	θ	$\theta+1$
1	1	0	$\theta+1$	θ
θ	θ	$\theta+1$	0	1
$\theta+1$	$\theta+1$	θ	1	0.

Note that the prime subfield $\mathbb{F}_2$ is in the upper left-hand corner. In general, the prime subfield $\mathbb{F}_p$ consists of the constant polynomials in $\mathbb{F}_p[\theta]$.

The multiplication table is as follows:

$\times$	0	1	θ	$\theta+1$
0	0	0	0	0
1	0	1	θ	$\theta+1$
θ	0	θ	$\theta+1$	1
$\theta+1$	0	$\theta+1$	1	θ.

Here is a sample computation needed in this table, remembering that all computations of the coefficients are being done modulo 2:

$$(\theta + 1)(\theta + 1) = \theta^2 + 2\theta + 1 \equiv \theta^2 + 1 \pmod{2}.$$

Now dividing this polynomial by $P(\theta) = \theta^2 + \theta + 1$ and taking the remainder, we get the answer $-\theta \equiv \theta \pmod{2}$, as recorded in the table.

Now the field $\mathbb{F}_9 = \mathbb{F}_{3^2}$:

Again $e = 2$, so we need a monic quadratic irreducible polynomial $P(\theta)$ over $\mathbb{F}_3$. In Section 10.7 we will see that there are three of these, and we pick $P(\theta) = \theta^2 + \theta + 2$. The addition table (which does not depend on the particular choice of $P(\theta)$) is:

+	0	1	2	θ	$\theta+1$	$\theta+2$	2θ	$2\theta+1$	$2\theta+2$
0	0	1	2	θ	$\theta+1$	$\theta+2$	2θ	$2\theta+1$	$2\theta+2$
1	1	2	0	$\theta+1$	$\theta+2$	θ	$2\theta+1$	$2\theta+2$	2θ
2	2	0	1	$\theta+2$	θ	$\theta+1$	$2\theta+2$	2θ	$2\theta+1$
θ	θ	$\theta+1$	$\theta+2$	2θ	$2\theta+1$	$2\theta+2$	0	1	2
$\theta+1$	$\theta+1$	$\theta+2$	θ	$2\theta+1$	$2\theta+2$	2θ	1	2	0
$\theta+2$	$\theta+2$	θ	$\theta+1$	$2\theta+2$	2θ	$2\theta+1$	2	0	1
2θ	2θ	$2\theta+1$	$2\theta+2$	0	1	2	θ	$\theta+1$	$\theta+2$
$2\theta+1$	$2\theta+1$	$2\theta+2$	2θ	1	2	0	$\theta+1$	$\theta+2$	θ
$2\theta+2$	$2\theta+2$	2θ	$2\theta+1$	2	0	1	$\theta+2$	θ	$\theta+1$.

Again, notice the prime subfield $\mathbb{F}_3$ in the upper left-hand corner.

Here is the multiplication table (which *does* depend on our choice of the monic irreducible polynomial $P(\theta)$):

×	0	1	2	θ	$\theta+1$	$\theta+2$	2θ	$2\theta+1$	$2\theta+2$
0	0	0	0	0	0	0	0	0	0
1	0	1	2	θ	$\theta+1$	$\theta+2$	2θ	$2\theta+1$	$2\theta+2$
2	0	2	1	2θ	$2\theta+2$	$2\theta+1$	θ	$\theta+2$	$\theta+1$
θ	0	θ	2θ	$2\theta+1$	1	$\theta+1$	$\theta+2$	$2\theta+2$	2
$\theta+1$	0	$\theta+1$	$2\theta+2$	1	$\theta+2$	2θ	2	θ	$2\theta+1$
$\theta+2$	0	$\theta+2$	$2\theta+1$	$\theta+1$	2θ	2	$2\theta+2$	1	θ
2θ	0	2θ	θ	$\theta+2$	2	$2\theta+2$	$2\theta+1$	$\theta+1$	1
$2\theta+1$	0	$2\theta+1$	$\theta+2$	$2\theta+2$	θ	1	$\theta+1$	2	2θ
$2\theta+2$	0	$2\theta+2$	$\theta+1$	2	$2\theta+1$	θ	1	2θ	$\theta+2$.

For a sample calculation in this table (again, all coefficient calculations are done modulo 3):

$$(2\theta + 2)(2\theta + 1) = 4\theta^2 + 6\theta + 2 \equiv \theta^2 + 2 \pmod{3}.$$

Dividing by $P(\theta) = \theta^2 + \theta + 2$ and taking the remainder gives us $-\theta \equiv 2\theta \pmod 3$, as recorded in the table.

One final remark is in order. Though we will not prove it here, it is a fact that the finite field $\mathbb{F}_{p^e}$ is *unique* of that order. That is, if F is a finite field of order p^e, then it has the exact same structure as $\mathbb{F}_{p^e}$. (In abstract algebra, it is said that F and $\mathbb{F}_{p^e}$ are "isomorphic."). For example, if we were to use some quite different method than the one we employed here to construct $\mathbb{F}_{p^e}$ (namely, using an irreducible polynomial of degree e over $\mathbb{F}_p$), the elements might be labeled differently, but the additive and multiplicative structures resulting from the two constructions would be identical.

10.5   The Multiplicative Structure of $\mathbb{F}_q$

One surprising fact about finite fields is that their multiplicative structure is as simple as possible, in a sense that we will now make precise. In order to do this, we need a little more language from abstract algebra. We know that in general a field F possesses both an addition and a multiplication which must obey certain rules, for example we know that every non-zero element of F must have a multiplicative inverse in F. In abstract algebra we say that the elements of F form a "group" under addition, and that the *non-zero* elements of F (often denoted F^*) form a group under multiplication. If we now focus in on the case $F = \mathbb{F}_q$ (where q is a prime power), we know that the additive group has order q and the multiplicative group has order $q - 1$. If $q = p^e$ (p prime), we also know that the *additive* structure is simply addition modulo p if $e = 1$ and is polynomial addition modulo p if $e > 1$. The *multiplicative* structure is multiplication modulo p when $e = 1$, but we saw in the previous section that when $e > 1$, multiplication seems more complicated. It turns out, however and as we said above, that even though *doing* multiplication is tricky, the underlying multiplicative structure is not complicated.

One of the fundamental and beautiful theorems of the finite group theory is Lagrange's Theorem, which says quite simply that *the order of a subgroup of a finite group must divide the order of the group.* Here we wish to apply this result to to a certain type of subgroup called a *cyclic* subgroup, as follows.

Suppose $a \neq 1$ is an element of $\mathbb{F}_q^*$ (which is a multiplicative group of order $q-1$) and consider the set $\{a, a^2, a^3, \ldots\}$ of powers of a. Since $\mathbb{F}_q^*$ is finite, eventually some power a^m will be the first to have the same value as a previous power a^n. But $a^m = a^n$ tells us that $a^{m-n} = 1$. Writing $k = m - n$, we say that k is the *order* of the element a in the multiplicative group $\mathbb{F}_q^*$; i.e., k is the smallest positive integer such that $a^k = 1$. Moreover, the set of k elements $\{1, a, a^2, \ldots, a^{k-1}\}$ form a subgroup of $\mathbb{F}_q^*$. This subgroup is called the *cyclic subgroup generated by* a, and is often denoted $< a >$. We may also denote the order of a by $|a|$.

Now applying Lagrange's Theorem, to the cyclic subgroup $< a >$ inside the group $\mathbb{F}_q^*$, we must have that k (the multiplicative order of a and hence of $< a >$) must divide $q-1$, say $q-1 = sk$ for some integer s. Hence we have that $a^{q-1} = a^{sk} = (a^k)^s = 1^s = 1$; that is, we have shown that for every element a of $\mathbb{F}_q^*$, $a^{q-1} = 1$, and the order $|a|$ of a divides $q - 1$.

We come now to a key definition: An element b of $\mathbb{F}_q^*$ is called *primitive* if its order $|b|$ equals $q-1$ (not just divides $q-1$). (Note: This idea was first introduced in Section 6.3 in the form of primitive "roots" of $\mathbb{Z}_p$ using Fermat's Theorem.) *We now observe that if at least one primitive element exists in $\mathbb{F}_q^*$, then $\mathbb{F}_q^*$ is itself a cyclic group.* We shall now prove this fact, showing that the multiplicative structure of every finite field is that of a cyclic group, which is the simplest structure a group can have, being generated by the powers of a single element.

We first need the following:

Lemma 10.4. *If x and y are elements of a finite multiplicative group, then the order $|xy|$ of their product xy is the least common multiple of the orders $|x|$ and $|y|$. In particular, if $|x|$ and $|y|$ are relatively prime, then $|xy| = |x||y|$.*

Proof. Let $d = \gcd(|x|, |y|)$. We observe that

$$(xy)^{(|x||y|)/d} = x^{|x|(|y|/d)} y^{|y|(|x|/d)} = 1^{y/d} 1^{x/d} = 1,$$

so $|xy|$ must divide $(|x||y|)/d$, which is the least common multiple of $|x|$ and $|y|$. We claim that in fact $|xy|$ cannot be smaller than this. Suppose $k > d$ divides $|x|$, then

$$(xy)^{(|x||y|)/k} = x^{(|x|/k)|y|}y^{|y|(|x|/k)} = x^{(|x|/k)|y|}1^{|x|/k} = x^{(|x|/k)|y|}.$$

Suppose that $x^{(|x|/k)|y|} = 1$, then $|x|$ must divide $(|x|/k)|y|$; i.e., there is an integer a such that $a|x| = (|x|/k)|y|$, and so $a = |y|/k$. But this is a contradiction since $k > d$ and k divides $|x|$, so k cannot divide $|y|$. A parallel argument can be made if $k > d$ divides $|y|$. Hence $x^{(|x|/k)|y|} \neq 1$, and our result follows. $\square$

Here then is our main result, whose statement is straight-forward but whose proof is a bit tricky.

Theorem 10.5. *The multiplicative group* $\mathbb{F}_q^*$ *of all non-zero elements of the finite field* $\mathbb{F}_q$ *is cyclic.*

Proof. The case where $q = 2$ is trivial so we assume that $q \geq 3$. For notational simplicity, we set $q - 1 = h > 1$. Suppose we have the prime factorization $h = \prod_{i=1}^{t} p_i^{r_i}$. For each i consider the polynomial $f_i(x) = x^{h/p_i} - 1$. This polynomial has degree $h/p_i < h$ and thus has at most h/p_i roots. Choose a_i, an element of $\mathbb{F}_q^*$ which is *not* a root of the polynomial f_i, so that $a_i^{h/p_i} \neq 1$. Let $b_i = a_i^{h/p_i^{r_i}}$ for each $i \leq t$.

We now show that the multiplicative order of b_i is $p_i^{r_i}$. By definition of b_i we have

$$b_i^{p_i^{r_i}} = (a_i^{h/p_i^{r_i}})^{p_i^{r_i}} = a_i^h = 1.$$

Thus the order of b_i must divide $p_i^{r_i}$ and so must be a power of the prime p_i. Assume that $b_i^{p_i^k} = 1$, for some $k < r_i$. Then we have

$$b_i^{p_i^{r_i-1}} = b_i^{(p_i^k p_i^{r_i-(k+1)})} = 1^{p_i^{r_i-(k+1)}} = 1.$$

But this is impossible, because $b_i^{p_i^{r_i-1}} = a^{h/p_i}$ and $a_i^{h/p_i} \neq 1$. Hence the order of b_i is exactly $p_i^{r_i}$, and we now apply Lemma 10.4 to the element $c = \prod_{i=1}^{t} b_i$ to get

$$|c| = |\prod_{i=1}^{t} b_i| = \prod_{i=1}^{t} |b_i| = \prod_{i=1}^{t} p_i^{r_1} = d = q - 1.$$

Thus c is a primitive element of $\mathbb{F}_q^*$, and the proof is complete. $\square$

Example 10.2. Let's illustrate the proof technique of Theorem 10.5 in the finite field $\mathbb{F}_{13}$, recalling that all arithmetic is done modulo 13. Using the notation in that proof, $q-1 = h = 12 = 2^2 \cdot 3$, so $p_1 = 2$, $r_1 = 2$, $p_2 = 3$ and $r_2 = 1$. The polynomials are $f_1(x) = x^{12/2} - 1 = x^6 - 1$ and $f_2(x) = x^{12/3} - 1 = x^4 - 1$.

We seek an element a_1 of $\mathbb{F}_{13}^*$ which is *not* a root of f_1. The element 5 will work since the powers of 5 modulo 13 are $\{5, 12, 8, 1\}$; i.e., 5 is a root of f_2, not f_1. So, we set $a_1 = 5$, and now $b_1 = a_1^{12/4} = 5^3 \equiv 8 \pmod{13}$. The proof then shows that $|8| = p_1^{r_1} = 2^2 = 4$. Skipping the details, we can likewise set $a_2 = 4$, $b_2 = 4^{12/3} = 4^4 \equiv 9 \pmod{13}$, so $|9| = 3^1 = 3$. Finally then, setting $c = b_1 b_2 = 8 \cdot 9 \equiv 7 \pmod{13}$, we get $|7| = |c| = |b_1||b_2| = 4 \cdot 3 = 12$, i.e., 7 is primitive in $\mathbb{F}_{13}^*$.

In fact we can check; here are the modulo 13 powers of 7:

$$\{7, 7^2 \ (\text{mod } 13), \ldots\} = \{7, 10, 5, 9, 11, 12, 6, 3, 8, 4, 2, 1\},$$

so 7 is indeed primitive in $\mathbb{F}_{13}$. We remark that in small finite fields primitive elements can be identified by trial and error, as we could have easily done here, but in larger fields that can be difficult. The proof of Theorem 10.5 gives us an algorithm which will always identify a primitive element by breaking the problem down into the cases of each prime dividing $q - 1$.

Finally, how many primitive elements does $\mathbb{F}_q$ have? We shall explore this is in Problem 10.15.

10.6   The Subfields of $\mathbb{F}_{p^n}$

We know that for all primes p, the prime field $\mathbb{F}_p$ is a subfield of the field $\mathbb{F}_{p^e}$ for all positive integer exponents e. Our question here is, What other subfields does $\mathbb{F}_{p^e}$ have? The answer turns out to be independent of p, depending only on the exponent e. Let's explore this question.

As we discussed in Section 10.5, the elements of any finite field form an additive group and the non-zero elements form a multiplicative group. Hence if K is a subfield of $\mathbb{F}_{p^e}$, then K is an additive subgroup of $\mathbb{F}_{p^e}$ and K^* (the set of non-zero elements of K) is a multiplicative subgroup of $\mathbb{F}_{p^e}^*$. Lagrange's Theorem (again, see Section 10.5) then tells us that the order $|K|$ of K must divide p^e and the order $|K^*|$ of K^* must divide $p^e - 1$.

We now know that $|K|$ must be p^d for some $1 \leq d \leq e$, which means that $|K^*| = p^d - 1$, and we must have that $p^d - 1$ divides $p^e - 1$. The following lemma does the trick.

Lemma 10.6. *For any positive integers x, e, and d, $x^d - 1$ divides $x^e - 1$ if and only if d divides e.*

Proof. Suppose d divides e, say $e = kd$, then

$$x^e - 1 = x^{kd} - 1 = (x^d)^k - 1 = (x^d - 1)((x^d)^{k-1} + (x^d)^{k-2} + \cdots + x^d + 1),$$

which says that $x^d - 1$ divides $x^e - 1$.

On the other hand, suppose that $x^d - 1$ divides $x^e - 1$. By the Division Algorithm $e = kd + r$ for some k and some $0 \leq r < d$. We have then

$$x^e - 1 = x^{kd} x^r - x^r + x^r - 1 = x^r (x^d - 1)((x^d)^{k-1} + \cdots + x^d + 1) + (x^r - 1).$$

Setting $N = x^r((x^d)^{k-1} + \cdots + x^d + 1)$, we have $x^e - 1 = N(x^d - 1) + (x^r - 1)$. By hypothesis and since $x^r - 1 < x^d - 1$, the Division Algorithm tells us that $x^r - 1 = 0$, so $r = 0$, and so d divides e. $\square$

We have now established our desired result.

Theorem 10.7. *For all primes p and all positive integers e and d, the subfields of the finite field $\mathbb{F}_{p^e}$ are exactly the finite fields $\mathbb{F}_{p^d}$ where d divides e.*

Example 10.3. For any prime p, the subfields of $\mathbb{F}_{p^{12}}$ are $\mathbb{F}_{p^6}$, $\mathbb{F}_{p^4}$, $\mathbb{F}_{p^3}$, $\mathbb{F}_{p^2}$, and the prime field $\mathbb{F}_p$. Below is the "lattice of subfields" of $\mathbb{F}_{p^{12}}$, where a line (or lines) between two fields indicates that the lower one is a subfield of the upper one.

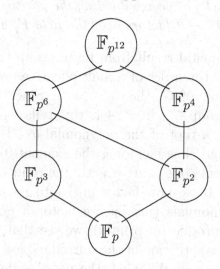

10.7 Counting Irreducible Polynomials over $\mathbb{F}_{p^n}$

In Section 10.4, while discussing a procedure for constructing the finite field $\mathbb{F}_{p^n}$, we asserted that, given any prime p and any positive integer exponent e, there always exists a monic irreducible polynomial $P(\theta)$ of degree n with coefficients in the prime field $\mathbb{F}_p$. In this section we not only show that this is true, but we also show that we can count exactly how many such polynomials there are. It turns out that the Möbius Inversion Formula (Theorem 8.9) comes into play here.

Let $N_q(n)$ denote the number of monic irreducible polynomials of degree n over the finite field F_q, i.e., with coefficients of the polynomial in the field F_q, where q is a prime power.

We need the following result:

Lemma 10.8. *Let T_n be the set of all monic irreducible polynomials over the field F_q of degree dividing the positive integer n. Then the polynomial $\theta^{q^n} - \theta$ factors over the field F_q as $\prod_{f \in T_n} f$.*

Proof. A fundamental result from finite group theory in abstract algebra is that if the order of a multiplicative group G is $|G|$ and if a is an element of G, then $a^{|G|} = 1$. Since the order of the multiplicative group $\mathbb{F}_{q^n}^*$ is $q^n - 1$, this tells us that if $a \neq 0$ is in $\mathbb{F}_{q^n}$, then a is a root of the polynomial $\theta^{q^n-1} - 1$. Multiplying through by θ (and thus including the element 0) we get that the q^n distinct elements of $\mathbb{F}_{q^n}$ are exactly the roots of the polynomial $g(\theta) = \theta^{q^n} - \theta$. If we now factor $g(\theta)$ into its product of monic irreducible polynomials (just as we factor a positive composite integer into its product of primes), we see that every element of $\mathbb{F}_{q^n}$ is a root of exactly one of these irreducibles.

A question now is, What are the possible degrees of these irreducible polynomials? The answer is: The degrees are numbers k which divide n. The basic idea is the following: The roots of a monic irreducible of degree k over $\mathbb{F}_q$ all come from the set of elements of the field $\mathbb{F}_{q^k}$ which do not lie in any proper subfields of $\mathbb{F}_{q^k}$ (such elements are called *primitive in* $\mathbb{F}_{q^k}$). However, all this is occurring inside the field $\mathbb{F}_{q^n}$, so $\mathbb{F}_{q^k}$ must be a subfield of $\mathbb{F}_{q^n}$, and in Theorem 10.7 we proved that k must be a divisor of n. (For example, looking at the subfield lattice at the end of the previous section, the roots of a 6-th degree irreducible polynomial over $\mathbb{F}_p$ would come from those elements of $\mathbb{F}_{p^6}$ which do not lie in either $\mathbb{F}_{p^3}$ or $\mathbb{F}_{p^2}$.) Since $g(\theta)$ factors into the product of all of these monic irreducibles, the proof is complete. $\square$

We now are able, using the Möbius Inversion Formula, to derive our formula for the function $N_q(n)$.

Theorem 10.9. *For any prime power q and any integer $n \geq 1$, the number of monic irreducible polynomials of degree n over the* *field* $\mathbb{F}_q$ *is given by:*

$$N_q(n) = \frac{1}{n} \sum_{d|n} \mu(d) q^{n/d},$$

where μ *denotes the Möbius function.*

Proof. We know that $x^{q^n} - x$ is the product of all monic irreducible polynomials over the field F_q of degree dividing n. By comparing degrees, we see for every positive integer n that

$$q^n = \sum_{d|n} dN_q(d);$$

that is, adding up, over all divisors d of n, the number of monic polynomials of degree d times their d distinct roots, we get the total number of elements q^n of $\mathbb{F}_{q^n}$.

We now apply the Möbius Inversion Formula (Theorem 8.9), where $f(n)$ in the theorem is q^n here and $F(n)$ is the summation $\sum_{d|n} dN_q(d)$. The inversion formula now yields the equation

$$nN_q(n) = \sum_{d|n} \mu(d) q^{n/d}.$$

The theorem follows by dividing both sides by n. $\square$

Example 10.4. In Section 10.4 we claimed that there is only one monic irreducible polynomial of degree 2 over $\mathbb{F}_2$ (namely $\theta^2 + \theta + 1$). Let's check using Theorem 10.9:

$$N_2(2) = \frac{\mu(1)2^{2/1} + \mu(2)2^{2/2}}{2} = \frac{4-2}{2} = 1.$$

Likewise, we claimed that there are three monic irreducible polynomials of degree 2 over $\mathbb{F}_3$. Checking this:

$$N_3(2) = \frac{\mu(1)3^{2/1} + \mu(2)3^{2/2}}{2} = \frac{9-3}{2} = 3.$$

Hence our claims were correct.

The Möbius function can seem mysterious, but in fact it is a numerical formulation of a basic principle in counting theory known as *inclusion-exclusion*. The idea is simple: In counting something, one may have over-counted some items ("inclusion"), so one needs to remove those now from the count ("exclusion"). A simple example is counting the number of elements in the union $A \cup B$ of two finite sets A and B which overlap. Computing $|A| + |B|$ has counted the intersection $A \cap B$ twice, so the correct count for $|A \cup B|$ is $|A| + |B| - |A \cap B|$. Let's look at an example of using Theorem 10.9 from this point of view.

Example 10.5. Look back at the diagram of the lattice of subfields of $\mathbb{F}_{p^{12}}$ at the end of Section 10.6. To count the number $N_p(12)$ of monic irreducible polynomials of degree 12 over $\mathbb{F}_p$, it is sufficient to count those elements which do not lie any proper subfields of $\mathbb{F}_{p^{12}}$ (these elements are exactly the primitive elements of $\mathbb{F}_{p^{12}}$, as defined in Section 10.5). Hence we must exclude the elements of $\mathbb{F}_{p^6}$ and $\mathbb{F}_{p^4}$, but in so doing we have excluded the elements of $\mathbb{F}_{p^2}$ twice, so we need to add them back in. So here's what we compute for the number of primitive elements of $\mathbb{F}_{p^{12}}$:

$$|\mathbb{F}_{p^{12}}| - |\mathbb{F}_{p^6}| - |\mathbb{F}_{p^4}| + |\mathbb{F}_{p^2}| = p^{12} - p^6 - p^4 + p^2,$$

and to get the number of polynomials we divide by 12, the number of roots of each polynomial of degree 12. Compare this now with the computation of $N_p(12)$ from Theorem 10.9:

$$\frac{\mu(1)p^{12} + \mu(2)p^6 + \mu(3)p^4 + \mu(4)p^3 + \mu(6)p^2 + \mu(12)p}{12} = \frac{p^{12} - p^6 - p^4 + p^2}{12}.$$

As a specific example, if we set $p = 2$, then the number $N_2(12)$ of monic irreducible polynomials of degree 12 over $\mathbb{F}_2$ is

$$\frac{2^{12} - 2^6 - 2^4 + 2^2}{12} = \frac{4096 - 64 - 16 + 4}{12} = \frac{4020}{12} = 335.$$

In Problem 10.18 we will ask the reader to use Theorem 10.9 to show that for each prime power q and for each positive integer $n \geq 1$ there is at least one monic irreducible polynomial of degree n over the field F_q; i.e., to show that $N_q(n)$ is always greater than 0. We note however that when searching for monic irreducible polynomials, especially of large degree, it is in general not easy to find or construct such a polynomial for given values of q and n. Though we saw in Example 10.5 that there are 335 monic irreducible polynomials of degree 12 over $\mathbb{F}_2$, identifying even one of them can be a challenge.

In general, polynomials (not just irreducible ones) play an extremely important role in the theory of finite fields. In the upcoming section we will see that every function from a finite field to itself can be represented by a polynomial. In fact it can be shown that finite fields are the only algebraic structures that have this property of guaranteed polynomial representation of arbitrary functions from the structures to themselves.

10.8   Lagrange Interpolation Formula

As just indicated, we now state and prove the fact that every function from a finite field $\mathbb{F}_q$ to itself can be represented as a polynomial over that field. By a function $f(x)$ "being represented by" a polynomial $P_f(x)$, we mean that for each element $b \in \mathbb{F}_q$, $P_f(b) = f(b)$. This result, which we state in the following form, is named after Joseph-Louis Lagrange (1736 - 1813).

Theorem 10.10. (*Lagrange Interpolation Formula*) *Every function* $f : \mathbb{F}_q \to \mathbb{F}_q$ *can be represented by a unique polynomial over* $\mathbb{F}_q$ *of degree at most* $q - 1$. *Moreover, such a polynomial is given by*

$$P_f(x) = \sum_{a \in \mathbb{F}_q} f(a)[1 - (x - a)^{q-1}].$$

Proof. Let f be a function from the field $\mathbb{F}_q$ to itself and let the polynomial $P_f(x)$ be as stated in the theorem. Recall that the non-zero elements of $\mathbb{F}_q$ form a multiplicative group of order $q-1$, and by group theory every such element c satisfies $c^{q-1} = 1$. Hence $(b - a)^{q-1}$ is equal to 1 if $a \neq b$ and is equal to 0 if $a = b$. Thus we can quickly check that $P_f(b) = f(b)$ for all $b \in \mathbb{F}_q$, so the polynomial $P_f(x)$ does indeed represent the function $f(x)$ over the finite field $\mathbb{F}_q$. The uniqueness of $P_f(x)$ follows from the facts that

(1) the number of possible functions from $\mathbb{F}_q$ to itself is q^q, and

(2) the number of polynomials over $\mathbb{F}_q$ of degree less than q is also q^q.

Since these two sets are in one-to-one correspondence, $P_f(x)$ uniquely represents the function $f(x)$. This completes the proof. $\square$

Example 10.6. Here we illustrate this result by applying it to two functions f and g from $\mathbb{F}_3$ to itself. First, let f be the identity function $f(x) = x$ on $\mathbb{F}_3$, and let us confirm that the polynomial $P_f(x)$ defined in Theorem 10.10 is indeed $P_f(x) = x$. Remembering that the coefficients are in $\mathbb{F}_3$, we have that $P_f(x)$ is

$$0[1 - (x - 0)^2] + 1[1 - (x - 1)^2] + 2[1 - (x - 2)^2] = 1 - (x^2 - 2x + 1)$$

$$+2 - 2(x^2 - 4x + 4) = -x^2 + 2x + 2 - 2x^2 + 8x - 8 = -3x^2 + 10x - 6 \equiv x$$

(mod 3), as expected.

Next suppose that g is the function on $\mathbb{F}_3$ defined by $g(0) = 1$, $g(1) = 2$ and $g(2) = 2$. A polynomial $P_g(x)$ to represent this function is not obvious, but Theorem 10.10 allows us to compute it. $P_g(x)$ is

$$1[1-(x-0)^2]+2[1-(x-1)^2]+2[1-(x-2)^2] = 1-x^2+2-2(x^2-2x+1)$$

$$+2-2(x^2-4x+4) = 1-x^2+2-2x^2+4x-2+2-2x^2+8x-8 = -5x^2+$$

$$12x - 5 \equiv x^2 + 1 \ (\text{mod } 3).$$

Let's check this now: $P_g(0) = 0^2 + 1 = 1$, $P_g(1) = 1^2 + 1 = 2$ and $P_g(2) = 2^2 + 1 \equiv 2 \ (\text{mod } 3)$, so we have indeed represented g by a polynomial, this one being of maximal degree 2.

We can extend the previous result to functions of more than one variable.

Theorem 10.11. *For any integer $n \geq 1$, let $f : \mathbb{F}_q^n \to \mathbb{F}_q$. Then the function f can be represented by a unique polynomial $P_f(x_1, \ldots, x_n)$ over $\mathbb{F}_q$ of degree at most $q - 1$ in each variable. Moreover, such a polynomial is given by*

$$P_f(x_1, \ldots, x_n) = \sum_{(a_1, \ldots a_n) \in F_q^n} f(a_1, \ldots, a_n) \prod_{1 \leq i \leq n} [1 - (x_i - a_i)^{q-1}].$$

We leave the proof of this result to the reader in Problem 10.22.

10.9 An Application to Latin and Sudoku Squares

In this section we provide a beautiful application of finite fields to the construction of sets of mutually orthogonal Latin squares. By a *Latin square of order n* is meant an $n \times n$ square with the property that each of the n distinct elements occurs exactly once in each row and in each column. A pair of Latin squares of order n is said to be *orthogonal* if when placing one square on top of the other, in the n^2 cells, each of the ordered pairs (i, j) with $1 \leq i \leq n, 1 \leq j \leq n$, occurs exactly once. Finally a set of $t \geq 2$ Latin squares of order n is *mutually orthogonal* if any two distinct squares in the set are orthogonal. It is known that the maximal size of a set of mutually orthogonal Latin squares of order n is $n - 1$. Such a maximal set of squares is said to be *complete*.

Example 10.7. Here is a pair of orthogonal Latin squares of order 3. The reader should check that they are in fact Latin squares, and that they are in fact orthogonal.

0	1	2
1	2	0
2	0	1

0	1	2
2	0	1
1	2	0

We note also that this is a complete set since $3 - 1 = 2$.

We shall now see that finite fields are very useful in constructing complete sets of mutually orthogonal Latin squares whenever the order q of the square is a prime power, since then we have the field $\mathbb{F}_q$ to utilize. The following construction is usually credited to Raj Chandra Bose (1901 - 1987), though he was not the first person to use finite fields to construct such complete sets of orthogonal Latin squares.

Theorem 10.12. *Suppose that q is a prime power and that a $q \times q$ matrix has the rows labeled by $\{x \in \mathbb{F}_q\}$ and the columns by $\{y \in \mathbb{F}_q\}$. Then for any $a \neq 0$ in $\mathbb{F}_q$, if the values in $\mathbb{F}_q$ of the linear polynomial $ax + y$ are placed in the locations (x, y), the resulting matrix is a Latin square of order q. Moreover, as a runs through all the non-zero elements of $\mathbb{F}_q$, we obtain a complete set of $q - 1$ mutually orthogonal squares.*

Proof. We emphasize that all arithmetic is being done in the field $\mathbb{F}_q$, so in particular every non-zero element has a multiplicative inverse, and so we can do division.

We first show that for each $a \neq 0$ in $\mathbb{F}_q$, the procedure stated in the theorem results in a Latin square of order q. To first prove that the square has distinct elements in each row, suppose that in row x, $ax + y_1 = ax + y_2$. Then subtracting ax from both sides we see that $y_1 = y_2$, so that row x must contain distinct elements. To show that column y also contains distinct elements, assume that we have $ax_1 + y = ax_2 + y$. Subtracting y from both sides, we obtain that $ax_1 = ax_2$ and dividing by a we have that $x_1 = x_2$. Column y thus contains distinct elements as well, and we have produced a Latin square of order q.

We now show that if $a \neq 0$ and $b \neq 0$ are distinct elements of $\mathbb{F}_q$, then the two polynomials $ax + y$ and $bx + y$ produce Latin squares which are orthogonal. To prove orthogonality of the two squares, assume that the two squares are not orthogonal so that there are two cells in the squares which have ordered pairs with the same values; that is, suppose that

$$ax_1 + y_1 = ax_2 + y_2$$

$$bx_1 + y_1 = bx_2 + y_2$$

both hold, then upon subtraction we have that $(a - b)x_1 = (a - b)x_2$, but since a and b are distinct, we can divide by $a - b$ to obtain $x_1 = x_2$, and so $y_1 = y_2$ also. Hence the two squares are indeed orthogonal, and the proof is complete. $\square$

Note that the two orthogonal Latin squares in Example 10.7 arise from the linear polynomials $x + y$ and $2x + y$ over the field

$\mathbb{F}_3$. We also note that for any prime power q, the square generated by the polynomial $x + y$ simply gives us the addition table of $\mathbb{F}_q$.

Example 10.8. Let's consider the case $q = 4$. Look back now in Example 10.1 at the addition table for $\mathbb{F}_4$ and note that it is a Latin square of order 4 generated by the polynomial $x + y$. Here then are, respectively, the Latin squares generated by $\theta x + y$ and $(\theta + 1)x + y$:

	0	1	θ	$\theta + 1$
0	0	1	θ	$\theta + 1$
1	θ	$\theta + 1$	0	1
θ	$\theta + 1$	θ	1	0
$\theta + 1$	1	0	$\theta + 1$	θ

and

	0	1	θ	$\theta + 1$
0	0	1	θ	$\theta + 1$
1	$\theta + 1$	θ	1	0
θ	1	0	$\theta + 1$	θ
$\theta + 1$	θ	$\theta + 1$	0	1.

We remark that in computing the values, we do sometimes produce θ^2. Recall that if we then divide θ^2 by the monic irreducible polynomial $\theta^2 + \theta + 1$, we get a remainder of $-\theta - 1 \equiv \theta + 1$ (mod 2), and we go from there. What we have then, by Theorem 10.12, is a complete set of 3 mutually orthogonal Latin squares of order 4.

One other remark is in order about these two Latin squares above which will lead us to our next topic. Notice that in both 4×4 squares, if we look at the four 2×2 sub-squares, we see that all four elements of $\mathbb{F}_4$ appear in each sub-square (but this is not true of the $\mathbb{F}_4$ addition table in Example 10.1). Hence these latter two squares are special, and we now take a look at this type of Latin square.

Sudoku squares are a very popular special case of Latin squares. A *Sudoku square* of order 9 is a 9×9 square, using the numbers

$1, 2, \ldots, 9$, so that each of the nine numbers occurs exactly once in each row, in each column, *and* in each of the nine 3×3 sub-squares. We now show that the finite field $\mathbb{F}_9$ can be used to construct such squares which are in fact mutually orthogonal by Theorem 10.12.

Just as before, consider the linear polynomials $ax + y$, with $a \in F_9$ and $a \neq 0$, but this time we also require that a not be 1 or 2 either. The reason for this is the following: Look back now at the addition table for $\mathbb{F}_9$ given in Example 10.1, which is generated by the case $a = 1$, and observe that *none* of the nine sub-squares contain all nine of the field elements. Hence this is a Latin square but is definitely not a Sudoku square. However, it turns out that for the other six polynomials ($\theta x + y$, etc.) we do indeed generate Sudoku squares, though we shall not prove this here. In fact it can be shown that those six squares are the maximum number of mutually orthogonal Sudoku squares of order nine.

Example 10.9. Below is the Latin square of order 9 generated by the polynomial $\theta x + y$, and we can check that we do have a Sudoku square as a result. As in Example 10.8, when we encounter θ^2, we need to divide by our chosen monic irreducible polynomial $\theta^2 + \theta + 2$ (again see Example 10.1) and take the remainder, which is $-\theta - 2 \equiv 2\theta + 1 \pmod 3$. Here is what we get after the numerous calculations:

	0	1	2	θ	$\theta+1$	$\theta+2$	2θ	$2\theta+1$	$2\theta+2$
0	0	1	2	θ	$\theta+1$	$\theta+2$	2θ	$2\theta+1$	$2\theta+2$
1	θ	$\theta+1$	$\theta+2$	2θ	$2\theta+1$	$2\theta+2$	0	1	2
2	2θ	$2\theta+1$	$2\theta+2$	0	1	2	θ	$\theta+1$	$\theta+2$
θ	$2\theta+1$	$2\theta+2$	2θ	1	2	0	$\theta+1$	$\theta+2$	θ
$\theta+1$	1	2	0	$\theta+1$	$\theta+2$	θ	$2\theta+1$	$2\theta+2$	2θ
$\theta+2$	$\theta+1$	$\theta+2$	θ	$2\theta+1$	$2\theta+2$	2θ	1	2	0
2θ	$\theta+2$	θ	$\theta+1$	$2\theta+2$	2θ	$2\theta+1$	2	0	1
$2\theta+1$	$2\theta+2$	2θ	$2\theta+1$	2	0	1	$\theta+2$	θ	$\theta+1$
$2\theta+2$	2	0	1	$\theta+2$	θ	$\theta+1$	$2\theta+2$	2θ	$2\theta+1$.

This square is a bit difficult to read, so let's translate it to a standard Sudoku square using the following (logical) assignments:

$$0 \to 1, 1 \to 2, 2 \to 3, \theta \to 4, \theta + 1 \to 5, \theta + 2 \to 6, 2\theta \to 7, 2\theta$$

$$+1 \to 8, 2\theta + 2 \to 9.$$

Now our square looks like this:

	1	2	3	4	5	6	7	8	9
1	1	2	3	4	5	6	7	8	9
2	4	5	6	7	8	9	1	2	3
3	7	8	9	1	2	3	4	5	6
4	8	9	7	2	3	1	5	6	4
5	2	3	1	5	6	4	8	9	7
6	5	6	4	8	9	7	2	3	1
7	6	4	5	9	7	8	3	1	2
8	9	7	8	3	1	2	6	4	5
9	3	1	2	6	4	5	9	7	8

and it can easily be checked to be a Sudoku square. We know that in addition five more squares can be produced using the other five primitive elements of $\mathbb{F}_9$ (i.e., $\theta + 1$, $\theta + 2$, 2θ, $2\theta + 1$, and $2\theta + 2$) as the coefficient of x.

For further study of this topic, see, for example, Laywine and Mullen [4].

10.10 Summary

In this chapter we discovered that beyond the sets $\mathbb{Z}_p$ (where p is prime), which we learned about in Chapter 3 and which are finite fields since all their non-zero elements possess multiplicative inverses, we are able to construct finite fields of order p^e where e can be greater than 1. We proved that for every prime p and every integer $e \geq 1$ there is a finite field $\mathbb{F}_{p^e}$ of order p^e (which is unique, though we did not prove that), and we proved that *every* finite field is one of these. We then

1. learned about the subfield structure of finite fields,
2. used the Möbius function to count monic irreducible polynomials of degree n over $\mathbb{F}_q$,
3. discovered that every function from $\mathbb{F}_q$ to itself can be represented by a polynomial over $\mathbb{F}_q$ of degree at most $q - 1$, and
4. used finite fields to construct Latin squares and their special case of Sudoku squares.

Finite fields have many practical applications, for example in information science using algebraic coding theory and in cryptology (as we learned in Chapter 6). See Mullen and Mummert [7] for an elementary treatment of finite fields and their applications. For a more in-depth treatment of finite fields, see the classic books by Lidl and Niederreiter [5] and [6], as well as the more recent volume edited by Mullen and Panario [8].

10.11   Solved Problems

Structure of $\mathbb{Z}_q$ Versus $\mathbb{F}_q$

10.1. Show that the set $\mathbb{Z}_4$ and the field $\mathbb{F}_4$ have different algebraic structure.

Solution:
In $\mathbb{Z}_4$, multiplication is done modulo 4, so $2 \cdot 2 \equiv 0 \pmod 4$, and so 2 cannot possess a multiplicative inverse in $\mathbb{Z}_4$. But in any field, every non-zero element does possess a multiplicative inverse.

The Order of a Finite Field

10.2. List the orders of all the finite fields of order less than or equal to 40.

Solution:
2, 3, 4, 5, 7, 8, 9, 11, 13, 16, 17, 19, 23, 25, 27, 29, 31, 32, 37.

Characteristic of a Field

10.3. (a) Show that for any prime p there are infinitely many different finite fields each with characteristic p.
(b) Show that for any positive integer $n \geq 1$, there are infinitely many finite fields F_{p^n} with different characteristics.

Solution:
For Part (a), fix a prime p and consider the infinite set of finite fields F_{p^n} for all n a positive integer.
For Part (b), fix n and consider the infinite set of finite fields F_{p^n} for all primes p (since there are infinitely many primes).

Constructing Finite Fields

10.4. In order to construct the field $\mathbb{F}_8$, we need to identify a monic irreducible polynomial of degree 3 over $\mathbb{F}_2$.
(a) According to Theorem 10.9, how many such polynomials are there?
(b) A cubic polynomial is irreducible if it cannot be divided by a linear polynomial. Use this to show that $P(\theta) = \theta^3 + \theta + 1$ is irreducible over $\mathbb{F}_2$.
(c) Use $P(\theta)$ from Part (b) to compute the product of the elements $\theta + 1$ and $\theta^2 + \theta$ in $\mathbb{F}_8$.

Solution:
(a) $N_2(3) = (\mu(1)2^3 + \mu(3)2^1)/3 = (8 - 2)/3 = 2$.
(b) There are only two linear polynomials over $\mathbb{F}_2$, θ and $\theta + 1$. Using the Division Algorithm for polynomials, we have $\theta^3 + \theta + 1 = (\theta)(\theta^2 + 1) + 1$, and $\theta^3 + \theta + 1 = (\theta + 1)(\theta^2 + \theta) + 1$, so in both cases the remainder is 1. Hence $\theta^3 + \theta + 1$ is irreducible.
(c) $(\theta + 1)(\theta^2 + \theta) = \theta^3 + \theta^2 + \theta + 1$, and now by the Division Algorithm

$$\theta^3 + \theta^2 + \theta + 1 = (\theta^3 + \theta + 1)(1) + \theta^2,$$

so the answer is the remainder θ^2.

The Order of Elements and Subfields

10.5. Write out the sets of all the powers modulo 11 of the 10 non-zero elements of $\mathbb{F}_{11}$ and list the order of each element.

Solution:
Each set below is of the form $\{a, a^2 \pmod{11}, a^3 \pmod{11}, \ldots\}$ until the value 1 is reached.

$\{1\}$, $|1| = 1$
$\{2, 4, 8, 5, 10, 9, 7, 3, 6, 1\}$, $|2| = 10$, primitive
$\{3, 9, 5, 4, 1\}$, $|3| = 5$
$\{4, 5, 9, 3, 1\}$, $|4| = 5$
$\{5, 3, 4, 9, 1\}$, $|5| = 5$
$\{6, 3, 7, 9, 10, 5, 8, 4, 2, 1\}$, $|6| = 10$, primitive
$\{7, 5, 2, 3, 10, 4, 6, 9, 8, 1\}$, $|7| = 10$, primitive
$\{8, 9, 6, 4, 10, 3, 2, 5, 7, 1\}$, $|8| = 10$, primitive
$\{9, 4, 3, 5; 1\}$, $|9| = 5$
$\{10, 1\}$, $|10| = 2$

10.6. For any prime power q, write down the orders of all the subfields of $\mathbb{F}_{q^{30}}$.

Solution:
The exponents in those orders are exactly the divisors of 30, so the orders are q, q^2, q^3, q^5, q^6, q^{10}, q^{15} and q^{30}.

Counting Monic Irreducible Polynomials

10.7. (a) How many monic irreducible polynomials of degree 3 are there over the field $\mathbb{F}_5$?
(b) How many monic irreducible polynomials of degree 6 are there over the field $\mathbb{F}_2$?
(c) How many monic irreducible polynomials of degree 4 are there over the field $\mathbb{F}_3$?

Solution:
(a) $N_5(3) = \mu(1)5^3 + \mu(3)5^1 = 125 - 5 = 120$.
(b) $N_2(6) = \mu(1)2^6 + \mu(2)2^3 + \mu(3)2^2 + \mu(6)2^1 = 64 - 8 - 4 + 2 = 54$.
(c) $N_3(4) = \mu(1)3^4 + \mu(2)3^2 + \mu(4)3^1 = 81 - 9 + 0 = 72$.

Representing Functions as Polynomials

10.8. Use Theorem 10.10 to find the unique polynomial $P_f(x)$ of degree less than 3 which represents the function $f(x)$ from $\mathbb{F}_3$ to itself defined by $f(0) = 1$, $f(1) = 0$ and $f(2) = 2$.

Solution:
By Theorem 10.10, we have that $P_f(x)$ is

$$1[1-(x-0)^2]+0[1-(x-1)^2]+2[1-(x-2)^2] = (1-x^2)+2-2(x^2-4x+4)$$

$$= 1 - x^2 + 2 - 2x^2 + 8x - 8 = -3x^2 + 8x + 1 \equiv 2x + 1 \pmod{3}.$$

Latin and Sudoku Squares

10.9. As in Theorem 10.12, we know that the Latin square of order 5 which is generated by the linear polynomial $x + y$ is simply the addition table of the field $\mathbb{F}_5$. Produce that addition table and verify that it is indeed a Latin square

Solution:

+	0	1	2	3	4
0	0	1	2	3	4
1	1	2	3	4	0
2	2	3	4	0	1
3	3	4	0	1	2
4	4	0	1	2	3

Permutation Polynomials

10.10. A polynomial f over a finite field $\mathbb{F}_q$ is a *permutation polynomial* if it is a $1-1$ onto mapping from F_q to itself. Show that a linear polynomial $ax + b$ with $a, b \in \mathbb{F}_q$, $a \neq 0$ is a permutation polynomial.

Solution:
To show that the polynomial $ax+b$ is $1-1$, assume that $ax_1+b = ax_2 + b$ so that upon subtraction we have $a(x_1 - x_2) = 0$. Then dividing by the non-zero value a we see that $x_1 = x_2$.
To show that the polynomial is onto, let $c \in \mathbb{F}_q$. Then with $x = (c - b)/a$, we have $ax + b = a((c - b)/a) + b = c$.

10.11. Determine the total number of permutation polynomials (see the previous problem) of degree at most $q - 1$ over the finite field $\mathbb{F}_q$.

Solution:

We know from Theorem 10.10 that polynomials over $\mathbb{F}_q$ of degree less than q are in one-to-one correspondence with all functions from $\mathbb{F}_q$ to itself. Starting with an element of the field $\mathbb{F}_q$, it can be mapped onto any of the q elements. Then taking another field element, in order to be $1 - 1$, it can be mapped onto any of the remaining $q - 1$ field elements. Continuing we see that there are a total of $(q)(q-1)(q-2)\cdots(2)(1) = q!$ permutation polynomials over the field F_q.

10.12   Supplementary Problems

Structure of $\mathbb{Z}_q$ Versus $\mathbb{F}_q$

10.12. Show that the set $\mathbb{Z}_9$ and the field $\mathbb{F}_9$ have different algebraic structures.

The Order of a Finite Field

10.13. List the orders of all the finite fields of order between 40 and 70.

Constructing Finite Fields

10.14. In order to construct the field $\mathbb{F}_{16}$, we need to identify a monic irreducible polynomial of degree 4 over $\mathbb{F}_2$.

(a) According to Theorem 10.9, how many such polynomials are there?

(b) The polynomial $P(\theta) = \theta^4 + \theta + 1$ is one of those you counted in Part (a). Use it to compute the product of the elements $\theta^2 + 1$ and $\theta^3 + 1$ in $\mathbb{F}_{16}$.

Counting Primitive Elements

10.15. In Problem 10.5 we saw that $\mathbb{F}_{11}^*$ has 4 primitive elements. Here is some more data:

Field	# Primitive Elements
$\mathbb{F}_{13}$	4
$\mathbb{F}_{16}$	8
$\mathbb{F}_{17}$	8
$\mathbb{F}_{19}$	6

(a) Based on this data, make a conjecture about how many primitive elements there are in $\mathbb{F}_q$. (Suggestion: Look at any one of the primitive elements in Problem 10.5. In what positions in the set of powers of that element do the other three primitive elements occur?)

(b) Prove that your conjecture is true. (Note: This is not an easy proof.)

Subfields

10.16. As was done in Section 10.6 for the field $\mathbb{F}_{p^{12}}$, draw the lattice of subfields of $\mathbb{F}_{p^{18}}$.

Counting Monic Irreducible Polynomials

10.17. (a) How many monic irreducible polynomials of degree 8 are there over the field $\mathbb{F}_2$?

(b) How many monic irreducible polynomials of degree 6 are there over the field $\mathbb{F}_3$?

10.18. Prove that for every finite field $\mathbb{F}_q$ and every positive integer n, there is at least one monic irreducible polynomial $P(\theta)$ of degree n over $\mathbb{F}_q$. (Suggestion: The leading term in the count is q^n. Each of the other terms is less than or equal to $q^{n/2}$, and there are fewer than n such terms. Go from there.)

Representing Functions as Polynomials

10.19. Use Theorem 10.10 to find the unique polynomial $P_f(x)$ of degree less than 3 which represents the function $f(x)$ from $\mathbb{F}_3$ to itself defined by $f(0) = 0$, $f(1) = 2$ and $f(2) = 2$.

10.20. Using the proof of Theorem 10.10 as a model, prove Theorem 10.11.

Permutation Polynomials

10.21. (a) Show that polynomial x^3 is a permutation polynomial (see Problem 10.10) of $\mathbb{F}_5$ to itself.
(b) Show that neither x^2 nor x^4 is a permutation polynomial of $\mathbb{F}_5$ to itself.
(c) More generally, prove that x^{2k} is not a permutation polynomial of $\mathbb{F}_q$ for all odd prime powers q and all $k > 0$.

Latin and Sudoku Squares

10.22. As in Theorem 10.12, use the linear polynomial $2x + 1$ to generate a Latin square of order 5, and verify that this square and the square in Problem 10.9 are orthogonal by examining the 25 ordered pairs.

10.23. Roll up your sleeves and use the polynomial $2\theta x + y$ to generate (as done with $\theta x + y$ in Example 10.9) a Sudoku square of order 9.

Answers to Selected Supplementary Problems

10.13. 41, 43, 47, 49, 53, 59, 61, 64, 67.

10.14. (a) 12 (b) $\theta^3 + \theta + 1$.

10.17. (a) 240 (b) 696.

10.19. $2x^2$.

10.22.

+	0	1	2	3	4
0	0	1	2	3	4
1	2	3	4	0	1
2	4	0	1	2	3
3	1	2	3	4	0
4	3	4	0	1	2

Chapter 11

Some Open Problems in Number Theory

11.1 Introduction

In this final chapter, we hope to inspire students (and instructors) to continue and extend their interest in mathematics by considering and studying a number of open problems that are directly related to the material in this text. Often students of mathematics believe that "everything has been solved" and that they are simply learning about those solutions, but that is not the case. In particular, as we shall see, although there are various mysteries about prime numbers which mathematicians would love to unravel, thus far they have been unable to do so.

Before embarking on our brief exploration of open problems in number theory, we bring up two problems which are *not* open but which have very different histories. The first of these is, quite simply, *Are there infinitely many primes?* This one was answered in the affirmative long ago by Euclid (around 250 BC) when he gave the first known proof of this fact, as follows: Suppose the set of primes is finite, i.e., suppose $\{2, 3, 5, \cdots, p_n\}$ is the complete set of n primes. Consider the number $N = (2 \cdot 3 \cdot 5 \cdots p_n) + 1$. If N were divisible by one of our primes p_k, then $1 = N - (2 \cdot 3 \cdot 5 \cdots p_n)$ would also be divisible by p_k, which is impossible. Hence N must be divisible some new prime, which contradicts the assertion that we had a list of all of the primes. We conclude then that the set of primes is infinite. There have been, in the ensuing centuries, many new proofs of this fact devised, but none as elegant as Euclid's original proof.

On the other hand, an example of a problem which remained unsolved for a very long time before finally being resolved is Fermat's Last Theorem, which we discussed in Chapter 9. The problem, first asked by Fermat in 1637, is, *If n is a positive integer greater than 2, are there are any solutions in positive integers of the Diophantine equation $x^n + y^n = z^n$?* In the years since special cases of the problem were resolved (for example, in that chapter, and with significant effort, we showed that the case $x^4 + y^4 = z^4$ has no such solutions), but it was not until 1995, about 360 years after Fermat made his conjecture, that Andrew Wiles (1953 -) was able to solve the general problem: For every $n > 2$, $x^n + y^n = z^n$ has no non-trivial integer solutions. So mathematicians chipped away at that problem for a long time before there was finally full success. Will this be what happens with all the open problems we list below, or will some of them never be resolved? Time will tell.

11.2 Open Problems

Open Problem 11.1. Mersenne Primes

In 1644 Marin Mersenne (1588 - 1648) conjectured that there were infinitely many primes of the form $M_n = 2^n - 1$ where n is a positive integer. Numbers of the form $2^n - 1$ are called *Mersenne numbers*, and if for some n this number M_n is prime, then M_n is called a *Mersenne prime*. Let's look at the value of the first ten Mersenne numbers and see if any patterns emerge:

n	1	2	3	4	5	6	7	8	9	10
$M_n = 2^n - 1$	1	3	7	15	31	63	127	255	511	1023

Which of the numbers M_n in this table are Mersenne primes?

Since 511 is divisible by 7 and 1023 is divisible by 3, we see that the numbers M_1, M_4, M_6, M_8, M_9, and M_{10} are all composite (i.e., not prime). This could lead us to make the following conjecture: *If n is composite, then M_n is also composite.* This conjecture turns out to be *true*, and we ask you to prove this in Problem 11.1. On the other hand, once we check that 127 is prime, we see that M_2, M_3, M_5, and M_7 are prime, which tempts us to make this conjecture: *If n is prime, then M_n is also prime.* However, this conjecture is immediately proven to be *false* since $M_{11} = 2047 = 23 \cdot 89$. It turns out, for example, that for the 15 primes p up to 50, eight of them satisfy that M_p is a Mersenne prime (besides the four already listed, they are M_{13}, M_{17}, M_{19}, and M_{31}) and the other seven are *not* Mersenne primes.

Mersenne's conjecture remains unsettled. As of the time of this writing there are only 51 Mersenne primes known, the largest being $2^{82,589,933} - 1$, which has 24,862,048 decimal digits. These very large Mersenne primes, quite obviously, have been discovered using computers. If you enjoy computer programming and might be interested in helping with the search for other Mersenne primes, you can do so by joining GIMPS (Great Internet Mersenne Prime Search) online. However, finding larger and larger Mersenne primes cannot settle our conjecture; that can be done only through a mathematical proof, and such a proof has not been discovered as of yet.

Open Problem 11.2. Perfect Numbers

First studied by Euclid, a *perfect number* is a positive integer n with the property that the sum of the divisors $\sigma(n) = 2n$ (or, said another way, n equals the sum of its *proper* divisors). As we observed in Example 8.4, 6 is a perfect number, and a second example is 28, which equals $1 + 2 + 4 + 7 + 14$. We may then ask the question, *Are there infinitely many perfect numbers?* In hoping to answer this question, Euclid proved the following lovely result:

Theorem 11.1. *If $2^m - 1$ is prime (i.e., if $2^m - 1$ is a Mersenne prime), then the positive integer $n = 2^{m-1}(2^m - 1)$ is a perfect number.*

We ask you to supply a proof in Problem 11.2. Recalling from Open Problem 11.1 that the first four Mersenne primes are M_2, M_3, M_5, and M_7, we use this formula to list four perfect numbers.

m	2	3	5	7
$2^{m-1}(2^m - 1)$	$2 \cdot 3 = 6$	$4 \cdot 7 = 28$	$16 \cdot 31 = 496$	$64 \cdot 127 = 8128$

Notice that just above we did not say the "*first* four perfect numbers" because Euclid's theorem does not say that *every* perfect number is of this form. However, much later Euler proved that all *even* perfect numbers are indeed of this form, so we have listed the first four *even* perfect numbers above. (We ask you to compute the fifth one in Problem 11.2.)

Hence there will be an infinite number of even perfect numbers if and only if there are an infinite number of Mersenne primes, and we know from Open Problem 11.1 that this is as yet unknown. We do know (again from Open Problem 11.1) that 51 even perfect numbers have been identified, the largest of which, discovered in 2018, has 49,724,095 decimal digits.

Finally, how about *odd* perfect numbers? As of now, no one has ever found an odd perfect number, but on the other hand no one has been able to prove odd perfect numbers cannot exist. If there *is* an odd perfect number, it is now known that it must be very large; currently it has been verified that any odd perfect number must be larger than 10^{1500}.

Open Problem 11.3. Fermat Primes

In analogy to examining expressions of the form $2^n - 1$ in Open Problem 11.1, what about expressions of the form $2^n + 1$. As we did previously, let's list the first ten of these and look for possible patterns:

n	1	2	3	4	5	6	7	8	9	10
$2^n + 1$	3	5	9	17	33	65	129	257	513	1025

This time the values arising from $n = 3, 5, 6, 7, 9, 10$ are all composite (since 129 and 513 are divisible by 3), whereas the values arising from $n = 1, 2, 4, 8$ are prime (we must check 257). This might lead us to make two conjectures. First, *If n is divisible by an odd prime, then $2^n + 1$ is composite.* This conjecture turns out to be *true*, and we ask you to prove it in Problem 11.3. The second conjecture is, *If n is a power of 2, then $2^n + 1$ is prime.* This conjecture turns out to be *false*, just as was the case for Mersenne numbers.

For each $k = 0, 1, 2, \ldots$, let $F_k = 2^{2^k} + 1$ (i.e., n in the above paragraph is a power of 2). A positive integer F_k of the form $2^{2^k} + 1$ is called a *Fermat number*, and it is a *Fermat prime* if it's prime. These numbers are so called and are denoted F_k in honor of Pierre de Fermat, who first studied them. Note that these numbers grow very quickly and so are quite large even for small values of k. We saw above that F_k is prime for $k = 0, 1, 2, 3$ (i.e., for $n = 1, 2, 4, 8$). It turns out that $F_4 = 65,357$ is also prime, *but $F_5 = 2^{2^5} + 1 = 2^{32} + 1 = 4,294,967,297$ is divisible by 641* and hence is composite. Currently it is not known if any other Fermat numbers F_k are prime for any larger values of k, though it is now known that $F_6 = 18,446,744,073,709,551,617$ is not prime. Nonetheless, the following question remains unanswered: *Are there infinitely many Fermat primes?* Again, as was the case with Mersenne primes, this question can only be answered by a mathematical argument, not by computer searches, and no such argument has been found thus far.

Open Problem 11.4. Formulas for the n-th Prime

If we list the primes in the usual order $2, 3, 5, 7, 11, 13, \ldots$, let p_n denote the n-th prime in the list. Can you find a formula for the n-th prime p_n; i.e., can you find a function f (polynomial, exponential, trigonometric, some combination of these, etc.) so that $f(n) = p_n$? As of today, no such function has ever been discovered.

There are however some polynomials which generate prime numbers for numerous values of n. A particularly amazing example is the polynomial $f(n) = n^2 + n + 41$. Again, let's look at some data by examining the values of this polynomial for n from 0 to 9:

n	0	1	2	3	4	5	6	7	8	9
$n^2 + n + 41$	41	43	47	53	61	71	83	97	113	131

All of these values are prime, and in fact it goes much further; $f(n)$ produces prime output for every n from 0 through 39. Finally at $n = 40$, $f(40) = 1681$, which is composite. In Problem 11.4 we ask you to show that for any positive constant c, the polynomial $g(n) = n^2 + n + c$ is a perfect square when $n = c - 1$, and hence is not prime. Going further, in that same problem we ask you to prove that *no* polynomial in $n = 0, 1, 2, \ldots$ can produce only prime numbers as output.

Open Problem 11.5. Gaps between Consecutive Primes

Given the n-th prime p_n, is there a formula to compute the next prime p_{n+1}? No such formula for the next prime has, at least so far, ever been found. This fact is not too surprising since the prime numbers distribute themselves very irregularly among the positive integers. Let us denote the number of composite numbers between p_n and p_{n+1} by g_n, and we call this the *gap* between these two consecutive primes. Clearly g_2 (i.e., the gap between 2 and 3) is the only case with $g_n = 0$. In the next Open Problem we take a close look at the case $g_n = 1$, i.e., the cases where $p_{n+1} = p_n + 2$, which are called *twin primes*. However, here let's examine the existence of wider gaps. Here are the first 100 numbers, with the primes highlighted:

1	2	3	4	5	6	7	8	9	10
11	12	**13**	14	15	16	**17**	18	**19**	·20
21	22	**23**	24	25	26	27	28	**29**	30
31	32	33	34	35	36	**37**	38	39	40
41	42	**43**	44	45	46	**47**	48	49	50
51	52	**53**	54	55	56	57	58	**59**	60
61	62	63	64	65	66	**67**	68	69	70
71	72	**73**	74	75	76	77	78	**79**	80
81	82	**83**	84	85	86	87	88	**89**	90
91	92	93	94	95	96	**97**	98	99	100

A close look at this array shows that it contains eight gaps of length 1 (i.e., eight pairs of twin primes), seven gaps of length 3, seven gaps of length 5, and one gap of length 7 (between 89 and 97). If we continued on, would wider gaps begin to appear? The answer is yes; in fact, we ask you to prove in Problem 11.5 that arbitrarily large gaps between consecutive primes exist. Hence, again, it is of little surprise that no one has found a formula for the number g_n, which would then be a formula for p_{n+1} given p_n since $p_{n+1} = p_n + g_n + 1$.

Open Problem 11.6. Twin Primes

Looking back at the array in the previous open problem, let us now concentrate on the smallest possible gaps between primes. As noted there, only 2 and 3 can have a gap of length 0 since from then on every other number is even, i.e., divisible by 2. Hence the smallest possible gap g_n becomes 1, meaning that $p_{n_1} = p_n + 2$. Two such primes are called *twin primes*. A look through that array reveals eight twin prime pairs below 100:

$$\{3,5\}, \{5,7\}, \{11,13\}, \{17,19\}, \{29,31\}, \{41,43\}, \{59,61\}, \{71,73\}.$$

As we move to larger numbers, the twin prime pairs begin to thin out, but they do keep occurring. It turns out that below 1000 there are 35 such pairs, and below 10,000 there are 205 such pairs. As we reminded you in the introduction, Euclid (and many others) proved that there are infinitely many primes, but the following question is unanswered as of now: *Are there infinitely many twin prime pairs?* The so-called *Twin Primes Conjecture* is that the answer is "yes."

The largest currently known twin prime pair is

$$2,996,863,034,895 \times 2^{1,290,000} \pm 1,$$

each of which has 388,342 decimal digits. But, again, interesting as this may be, it tells us nothing about the answer to our main question.

We now make a short visit to the area of mathematics known as "analytic number theory," which means, basically, using calculus techniques to get answers about number theory problems. Before returning to twin primes, we look at two questions simply involving integers. Those questions are:

1. Does the "harmonic series" $\sum_{n=1}^{\infty} 1/n$ converge or diverge?
2. Does the "1 over n-squared series" $\sum_{n=1}^{\infty} 1/n^2$ converge or diverge?

We can answer both of these questions using the basic calculus of "improper" integrals. We show here that the answer to Question 1 is "diverge," and in Problem 11.6 Part (b) we ask you to show that the answer to Question 2 is "converge." By looking at the area under the curve $1/x$ (from $x = 1$ on) compared to the sum of the areas of rectangles of height $1/n$ and width 1 (from $n = 1$ on), we see that

$$\sum_{n=1}^{\infty} 1/n > \int_{1}^{\infty} 1/x = \lim_{x \to \infty} \ln(x) = \infty;$$

i.e., the harmonic series diverges.

The important point for us here is that the sum of a set of real numbers can converge or diverge, but if a series converges that does *not* indicate that the set in the series is finite. Obviously the set of numbers $\{1/1, 1/4, 1/9, \cdots\}$ is infinite, but its sum is finite. In fact, it turns out that its sum is $\pi^2/6$.

Returning then to prime numbers in general and twin primes in particular, it has been proved that the series $\sum 1/p$ taken over all primes *diverges*, but the series $\sum 1/p$ taken only over twin primes *converges*. But as we saw above, this convergence does *not* imply that the set of twin prime pairs is a finite set. The Twin Primes Conjecture remains unresolved.

Finally, we remark that we can easily generalize the Twin Primes Conjecture. In the previous open problem we counted below 100 seven gaps of length 3, seven gaps of length 5, and one gap of length 7. Are there infinitely many of each of these? In general we can ask, Given an odd number k, are there infinitely many gaps between primes of length k? It should be no surprise that this question has not been answered for any value of k.

Open Problem 11.7. The Goldbach Conjecture

We learned in Chapter 2 that the prime numbers are the "building blocks of the multiplicative structure of the integers $\mathbb{Z}$" in that every integer $n \geq 2$ can be written uniquely (except for order) as a product of prime powers. But what role do the primes play in the *additive* structure of $\mathbb{Z}$?

Given an even integer $n \geq 4$, can n be written in the form $q_1 + q_2$ where q_1 and q_2 are both primes? Looking at some small cases, we see that $4 = 2 + 2$, $6 = 3 + 3$, $8 = 3 + 5$ and $10 = 5 + 5 = 3 + 7$. Starting with $n = 10$, there may well be more than one way to write n as the sum of two primes.

The *Goldbach Conjecture*, first stated by Christian Goldbach (1690 - 1764) in 1742, postulates that any even integer $n \geq 4$ can be written as the sum of two primes. This conjecture has now been verified by computer for very large even numbers, but thus far no one has proved that it holds for *all* even numbers.

Though no successful approaches to this problem have been found, the seemingly most promising approach has been to use analytic number theory, which we introduced briefly in Open Problem 11.6. The idea is to use calculus techniques to get an estimate of the number $N_2(n)$ of representations of n (an even integer) as a sum of two primes. If one could show that for all "sufficiently large" n (i.e., all n greater than some fixed n_0) $N_2(n)$ is greater than 0, then that would solve the problem for all but a finite number of positive integers. Hopefully then a computer could check the remaining finite number of integers below n_0. Unfortunately, even this approach has not been successful thus far in settling the Goldbach Conjecture.

If we can't solve the Goldbach Conjecture, perhaps we can prove that every *odd* number $n \geq 7$ is a sum of *three* primes. For example, $7 = 2 + 2 + 3$, $9 = 3 + 3 + 3$, $11 = 2 + 2 + 7 = 3 + 3 + 5$, and so on. Well, this problem did succumb to analytic techniques when the Russian mathematician Ivan Vinogradov (1891 - 1983) proved in 1934 that every sufficiently large odd integer n is a sum of three primes by showing that for such n, the number of "3-primes representations" $N_3(n)$ is greater than 0. So Vinogradov solved the problem for all but a finite number of positive integers, but his value n_0 is so large that even modern computers have not been able to fill in the additional cases. Nonetheless, solving the "3-Primes Problem" for sufficiently large integers was a major achievement. Still, the answer to the Goldbach Conjecture remains illusive.

Open Problem 11.8. The Number of Primes below n

We saw in Open Problem 11.5 that there are 25 primes below 100. It is also true that there are 168 primes below 1,000 and 1229 primes below 10,000. In general, we denote the function which, given a positive integer n, counts the number of primes below (or equal to) n by $\pi(n)$. (This is not to be confused with the famous number $\pi = 3.14159\ldots$, which measures the ratio of the circumference of a circle to its diameter.) So $\pi(100) = 25$, etc. *It is an open problem to find a formula for $\pi(n)$.*

That said, number theorists know quite a bit about *approximating* the function $\pi(n)$. After literally centuries of work, the following powerful theorem was finally proved in 1898 by two mathematicians, Jacques Hadamard and Charles Jean de la Vallee Poussin, working independently.

Theorem 11.2. (*The Prime Number Theorem - Version 1*) $\pi(n)$ *is approximated by* $n/\ln(n)$, *where* ln *is the natural logarithmic function. In fact, as n goes to infinity, the ratio* $\pi(n)/(n\ln(n))$ *approaches* 1.

Let's look at some data (decimal numbers are to five significant figures).

n	$\pi(n)$	$n/\ln(n)$	$\pi(n)/(n\ln(n))$
100	25	21.715	1.1513
1,000	168	144.77	1.1603
10,000	1229	1085.7	1.1320
100,000	9592	8685.9	1.1043
1,000,000	78498	72382	1.0845

Another way to express what this theorem says is that "in the vicinity" of the number N, the *density* of primes there is about one out of every $\ln(N)$ integers. So, for example, near 1,000, about one out of every $\ln(1,000) \approx 7$ integers is prime, whereas near 1,000,000, about one out of every $\ln(1,000,000) \approx 14$ integers is prime.

If we focus on the density of primes, another possible way of estimating $\pi(n)$ is to use analytic number theory, i.e., use calculus. If the density of primes "near" any real number x is approximated by $1/\ln(x)$, we can "add up" (i.e., *integrate*) these densities over the interval from 2 to n. That is, we can use that integral as follows (we state this version a little more succinctly).

Theorem 11.3. (*The Prime Number Theorem - Version 2*)

$$\pi(n) \approx \int_2^n 1/\ln(x)dx.$$

Moreover

$$\lim_{n \to \infty} \frac{\pi(n)}{\int_2^n 1/\ln(x)dx} = 1.$$

Let's add some data from this version to our chart above. Here we denote (as is standard) the integral $\int_2^n 1/\ln(x)dx$ by $Li(n)$ ("logarithmic integral").

n	$\pi(n)$	$n/\ln(n)$	$\pi(n)/(n/\ln(n))$	$Li(n)$	$\pi(n)/Li(n)$
100	25	21.715	1.1513	30.126	0.8298
1,000	168	144.77	1.1603	177.61	0.9459
10,000	1229	1085.7	1.1320	1236.1	0.9864
100,000	9592	8685.9	1.1043	9629.8	0.9961
1,000,000	78498	72382	1.0845	78629	0.9984

We note that the logarithmic integral seems to provide estimates which are closer to the actual values of $\pi(n)$. For example, at $n = 1,000,000$, the ratio $\pi(n)/(n/\ln(n))$ is "off" by about 8% whereas the ratio $\pi(n)/Li(n)$ is off by well less than 1%.

One note of interest in this data is that $n/\ln(n)$ seems to consistently underestimate the actual $\pi(n)$, whereas $Li(n)$ seems to overestimate $\pi(n)$. Whether this remains true for $n/\ln(n)$ as we go up is not clear, *but* in 1914 the English mathematician Dudley Edward Littlewood (1885 - 1977) proved that the quantity $\pi(n) - Li(n)$ changes signs infinitely often as n goes to infinity.

Finally, we ask you in Problem 11.8 to use calculus to show, without reference to $\pi(n)$ or the Prime Number Theorem, that

$$\lim_{n \to \infty} \frac{n/\ln(n)}{Li(n)} = 1.$$

Open Problem 11.9. The Riemann Hypothesis

This final open problem involves material somewhat more ·advanced than our previous ones and may not appear to have much to do with number theory. However, as we shall try to explain, it is of central importance in the theory of how the prime numbers are distributed within the integers $\mathbb{Z}$. We saw in Open Problems 11.5 and 11.6 how randomly the primes seem to be distributed, with very small and very large gaps between consecutive ones. If the conjecture (or "hypothesis") we encounter here were to be affirmed, that would shed much light on this distribution.

In Open Problem 11.6 we encountered the two series $\sum_{n=1}^{\infty} 1/n$ (the "harmonic series") and $\sum_{n=1}^{\infty} 1/n^2$ (the "1 over n-squared series"). We could generalize these two series into an infinite collection of series $f(r)$ by allowing the exponent on the n to be any real number r, so $f(r) = \sum_{n=1}^{\infty} 1/n^r$. We and you showed using comparison with improper integrals that $f(1)$ diverges and $f(2)$ converges. Generalizing, one can see that $f(r)$ diverges if $r \leq 1$ and converges if $r > 1$.

We take the final step now by allowing the exponent on n to be a complex number $s = r + ti$ where r and t are real numbers and $i = \sqrt{-1}$. The real number r is called the *real part* of s and is often denoted $Re(s)$. Now, the *Riemann zeta-function*, named for Bernhard Riemann (1826 - 1866), is defined by

$$\zeta(s) = \sum_{n=1}^{\infty} 1/n^s \text{ for all } s \text{ with } Re(s) > 1.$$

The complex numbers are normally pictured as the *complex plane*, where the horizontal axis represents the real part and the vertical axis the imaginary part of each complex number. Hence the Riemann zeta-function is initially defined only on the "open half plane" to the right of the vertical line $\{s | Re(s) = 1\}$. Without going into details, the function can now be "analytically continued" to the rest of the complex plane so that it is defined at every point except $(1,0)$, where, as we know, $\sum 1/n$ diverges (it is said to have a unique *pole* at $(1,0)$).

The connection of the Riemann zeta-function to the distribution of primes is revealed in the following identity, which was first established by Euler. We assume here that $Re(s) > 1$ and that p_i

is the i-th prime.

$$\zeta(s) = \sum_{n=1}^{\infty} 1/n^s = \prod_{i=1}^{\infty} \frac{1}{1 - 1/p_i^s} \, .$$

We ask you to work through the proof of this identity in Problem 11.9, which follows from unique factorization into prime powers (i.e., the Fundamental Theorem of Arithmetic) and some algebra.

We're getting to the hypothesis. The question is, Where does $\zeta(s)$ have the value 0. It turns out that it has what are called "trivial" zeros at the points $(-2, 0)$, $(-4, 0)$, $(-6, 0)$, etc., but the zeros of interest (the "non-trivial" zeros) all lie in what's called the *critical strip*, which is the vertical strip $\{s|0 < Re(s) < 1$. But Riemann suggested that those zeros are not just in the critical strip but actually all lie on the *critical line* at the middle of the strip, as follows:

The Riemann Hypothesis:
All non-trivial zeros of $\zeta(s)$ lie on the vertical line $\{s|Re(s) = 1/2\}$.

At this point many thousands of zeros of $\zeta(s)$ have been identified by computers, every one of which does indeed lie on the vertical line $\{s|Re(s) = 1/2\}$, but no one has succeeded in proving that all such zeros must be so located.

That such an odd sounding hypothesis might have a profound effect on our knowledge of how the prime numbers are distributed is hinted at by Euler's identity above, and it turns out that numerous results in number theory can either be improved or proved at all if the hypothesis is established. Let us give a single example. In Open Problem 11.7 we observed that the "3-primes problem" lacks a complete solution. However, the following was proved in the 1990's:

Theorem 11.4. *If the Riemann Hypothesis holds, then every odd number $n \geq 7$ can be written as a sum of three prime numbers.*

The proof uses the Riemann Hypothesis to drastically reduce Vinogradov's number n_0, and as a result the remaining finite number of cases could be verified using computation. Unfortunately, the Goldbach Conjecture does not appear to be helped by the Riemann Hypothesis.

11.3 Summary

As we come to the end of our journey through elementary number theory, we hope that encountering this set of open problems might encourage you the reader "to think like a mathematician" and be on the lookout for other problems, some easy, some perhaps hard or even very hard, to ponder and hopefully make progress on. Number theory is an exciting and fun area of mathematics since most of its settled and unsettled problems are easy to state but often hard to solve. We hope then that you will continue this journey as you move beyond this text.

11.4 Problems

(Note: Unlike all the other chapters in this text, in this chapter we have a single section of problems without solutions, though suggestions are given.)

Mersenne Primes

11.1. Prove the following result: If n is a composite positive integer, then $2^n - 1$ is also composite. (Suggestion: If $n = ab$ with a and b both positive, show that $(2^a)^b - 1$ factors into $2^a - 1$ times another positive factor which you can find using long division.)

Perfect Numbers

11.2. Prove Theorem 11.1, first proved by Euclid. (Suggestion: Make use of the σ-function, which we studied in Chapter 8, which (recall) adds up the divisors of an integer. Since we assume that $2^m - 1$ is prime, it is easy to compute $\sigma(2^m - 1)$. Also compute

$\sigma(2^{m-1})$, and since one of these arguments is odd and the other is a power 2, they must be relatively prime and so we can use the fact that σ is multiplicative.)

Fermat Primes

11.3. Prove that if n is divisible by an odd number b, then 2^n+1 is composite. (Suggestion: This proof is similar to the one in Problem 11.1. Write $n = ab$ and show how to factor $(2^a)^b+1$, using the fact that b is odd.)

Formulas for the n-th Prime

11.4. (a) Show that the polynomial $g(n) = n^2 + n + c$, where c is a positive integer, is a perfect square when $n = c - 1$, and hence is not prime.

(b) Much more generally, prove that the polynomial of positive degree k

$$h(n) = a_k n^k + a_{k-1} n^{k-1} + \cdots + a_1 n + a_0,$$

where each coefficient a_i is a non-negative integer, cannot have prime values for all $n = 0, 1, 2, \ldots$.

Gaps between Consecutive Primes

11.5. (a) Prove that given any positive integer m, there exists a sequence of at least m consecutive composite numbers (or, in the notation of Open Problem 11.5, for some n, $g_n \geq m$). (Suggestion: Consider the m consecutive numbers

$$(m + 1)! + 2, (m + 1)! + 3, \ldots, (m + 1)! + (m + 1).)$$

(b) In the array in Open Problem 11.5, the largest gap between consecutive primes below 100 has length 7 (between 89 and 97). In contrast, write down the seven consecutive composite numbers guaranteed by your proof of Part (a).

Twin Primes

11.6. (a) Identify the first twin prime pair above 100. Support your answer.

(b) By comparing it to an improper integral, prove that the series $\sum_{n=1}^{\infty} 1/n^2$ converges. (Suggestion: To get the needed relationship with the improper integral, you probably want to work with the series $\sum_{n=2}^{\infty} 1/n^2$. If that series converges, then so does the one with one more term. In fact, your proof should allow you to conclude that $\sum_{n=1}^{\infty} 1/n^2$ converges to a number between 1 and 2, which is true of its actual value, $\pi^2/6$.)

The Goldbach Conjecture

11.7. (a) Write each even number from 12 to 20 as a sum of two primes in as many ways as possible.

(b) Show that the number 100 can be written in six different ways as the sum of two primes.

(c) Show that the number 51 can be written in 15 different ways as the sum of three primes.

The Number of Primes below n

11.8. (a) What is the approximate density of primes near $n = 500,000$? (That is, about one out of every how many numbers near n is prime?)

(b) The two versions of the Prime Number Theorem say that both $n/\ln(n)$ and the logarithmic integral $Li(n) = \int_2^n 1/\ln(x)dx$ are "asymptotically equal" to $\pi(n)$, which would say that they are asymptotically equal to each other. We ask you to use calculus to show this directly.

(i) Using integration by parts, show that

$$\int_2^n \frac{1}{\ln(x)}dx = \frac{n}{\ln(n)} - \frac{2}{\ln(2)} + \int_2^n \frac{1}{(\ln(x))^2}dx.$$

(ii) Observe that the first term on the right above is $n/\ln(n)$. The second term is essentially negligible, but what about the integral in the third term? Now use l'Hopital's Rule and the Fundamental Theorem of Calculus to show that

$$\lim_{n\to\infty} \frac{\int_2^n 1/(\ln(x))^2 dx}{\int_2^n 1/\ln(x)dx} = 0.$$

Hence as n goes to infinity, that integral on the right also becomes negligible in comparison to the original integral. We can conclude then that

$$\lim_{n\to\infty} \frac{n/\ln(n)}{\int_2^n 1/\ln(x)dx} = \lim_{n\to\infty} \frac{n/\ln(n)}{Li(n)} = 1.$$

The Riemann Hypothesis

11.9. We wish to prove Euler's identity

$$\sum_{n=1}^{\infty} 1/n^s = \prod_{i=1}^{\infty} \frac{1}{1 - 1/p_i^s}$$

provided that s is a complex number with $Re(s) > 1$ and each p_i is prime. Let's use the following steps, each of which you should check carefully:

1. Let p be any prime. Show that for any positive integer k,

$$(1 - (1/p^s)^k) = (1 - 1/p^s)(1 + (1/p^s) + (1/p^s)^2 + \cdots + (1/p^s)^{k+1}).$$

2. Since $|1/p^s| < 1$, $\lim_{k\to\infty}|1/p^s|^k = 0$. Hence as $k \to \infty$ the left-hand side of the above equation becomes 1, and we get

$$\frac{1}{1 - 1/p^s} = 1 + (1/p^s) + (1/p^s)^2 + (1/p^s)^3 \cdots .$$

3. Now multiply this last equation over all primes p_i:

$$\prod_{i=1}^{\infty} \frac{1}{1 - 1/p_i^s} = \prod_{i=1}^{\infty}(1 + (1/p_i^s) + (1/p_i^s)^2 + (1/p_i^s)^3 \cdots).$$

4. Now apply the Fundamental Theorem of Arithmetic to the right hand side of this equation, which is

$$(1+1/2^s+1/4^s+\cdots)(1+1/3^s+1/9^s+\cdots)(1+1/5^s+1/25^s+\cdots)\cdots .$$

Verify that when this infinite product gets multiplied out, we get

$$1 + 1/2^s + 1/3^s + 1/4^s + 1/5^s + 1/6^s + \cdots = \sum_{n=1}^{\infty} 1/n^s,$$

and we are done.

Appendix A

Mathematical Induction

The principle of mathematical induction provides a very powerful method which can sometimes be used to prove that a certain property holds for all positive integers from some starting point on. For example, we show below, for all positive integers $n \geq 1$, that

$$1 + 2 + \cdots + n = n(n+1)/2,$$

i.e., the sum of the first n positive integers is $n(n+1)/2$.

We now state the principle of mathematical induction.

Theorem A.1. (*Mathematical Induction*) *Let $P(n)$ be a mathematical statement about the positive integer n. Assume that $P(1)$ holds and whenever $P(k)$ holds, so does $P(k+1)$. Then the statement $P(n)$ holds for all positive integers n.*

The base case in Theorem A.1 is the case $n = 1$, but we can start induction from any positive integer n_0, and if the induction step works, i.e., if the $n = k$ case implies the $n = k + 1$ case, then the property will hold for all positive integers $n \geq n_0$.

Example A.1. We now prove the above statement regarding the sum of the first n positive integers. Let $P(n)$ be the statement that for each positive integer n

$$1 + 2 + \cdots + n = n(n+1)/2.$$

DOI: 10.1201/9781003193111-A

We must first establish the base case. For $n = 1$ the sum contains just one term, namely 1, which of course equals $1(1+1)/2$, so the base case holds.

We now move on to the induction step. We *assume* that the statement $P(k)$ holds for any positive integer k; i.e., that

$$1 + 2 + \cdots + k = k(k+1)/2.$$

We now use this equality to prove that the statement $P(k+1)$ holds. In the following, the second equal sign is where we use our inductive assumption and the final two equal signs are due to algebraic manipulation.

$$1 + 2 + \cdots + k + (k+1) = (1 + 2 + \cdots + k) + (k+1)$$

$$= k(k+1)/2 + (k+1) = (k+1)(k/2 + 1) = (k+1)(k+2)/2,$$

which is the statement $P(k+1)$. Thus by Theorem A.1 our statement $P(n)$ holds for each positive integer $n \geq 1$, and we are done.

For example, the sum of the first 100 positive integers is seen to be

$$100(101)/2 = 50(101) = 5050.$$

Example A.2. As another example to illustrate the principle of mathematical induction, we now prove that for each positive integer n, the expression $2^{2n} - 1$ is divisible by 3.

For the base case $n = 1$, we have $2^2 - 1 = 3$, so the desired property holds. Though it is not needed for the proof, it is often helpful to look at a few cases just above the base case. So, for $n = 2$ we have $2^4 - 1 = 15$, and for $n = 3$ we have $2^6 - 1 = 63$, both of which are divisible by 3.

Now for the induction hypothesis we *assume* that $2^{2k} - 1 = 3M$ for some positive integer M. We have then

$$2^{2(k+1)} - 1 = 2^{2k+2} - 1 = 4 \cdot 2^{2k} - 1 = 4 \cdot 2^{2k} - 4 + 3$$

$$= 4(2^{2k} - 1) + 3 = 4(3M) + 3 = 3(4M + 1).$$

Thus given that 3 divides $2^{2k}-1$, we get that 3 divides $2^{2(k+1)}-1$, so the proof is complete. Be sure to identify in the sequence of equalities exactly where the induction hypothesis is employed. Every other equal sign in the sequence is by algebraic manipulation.

For another illustration of a proof using mathematical induction, we consider the following example which deals with the number of subsets of a given finite set.

Example A.3. We show by induction that if a set A has exactly n elements, then the power set of A, i.e., the set of all subsets of A, contains exactly 2^n subsets.

Let us check a couple of base cases. If A is the empty set (i.e., if $n = 0$), then the power set of A is just A itself, and $2^0 = 1$. If $n = 1$, assume that the set $A = \{a_1\}$. The subsets of A are the empty set and the set A itself, and $2^1 = 2$, so the result is true for $n = 1$. Going one step further, if $A = \{a_1, a_2\}$, then the subsets are the empty set, $\{a_1\}$, $\{a_2\}$ and A itself, and $2^2 = 4$ is correct.

We now move to the inductive step. Let B be a set with $k + 1$ elements and suppose b_0 is some fixed element of B. We define the set A to be the set $B - \{b_0\}$; that is, A is the set B with the single element b_0 removed. Hence A has k elements, and so by the induction hypothesis A contains 2^k subsets. Let us label those subsets as $T_1, T_2, \ldots, T_{2^k}$. Now suppose that S is a subset of B, then either b_0 is an element of S or it is not an element of S. In the latter case, S is a subset of A, i.e., $S = T_i$ for some i. In the former case, $S - \{b_0\}$ (i.e., S with b_0 removed) is a subset of A, so now $S = T_j \cup \{b_0\}$ for some j. Because all of the subsets listed here are distinct, and since S is of the form of one of them, the total number of subsets S of B, by our inductive hypothesis, is $2^k + 2^k = 2(2^k) = 2^{k+1}$, as we had hoped to prove.

We now briefly describe a second form of mathematical induction, often called "strong induction." Again $P(n)$ will denote a mathematical statement about the positive integer n. We first check that $P(1)$ holds (or that $P(n_0)$ holds for some positive integer n_0) to get the induction started. We then assume that the statement $P(m)$ holds for *all* $m \leq k$ and show that $P(k+1)$ also holds. Note that in the earlier form of induction, to prove that $P(k+1)$ is true, we only used that fact that the previous case $P(k)$ was true. In the strong form of induction, we assume that the statement holds for $m = k$, and that it *also* holds for each value $m < k$. The reason for the name strong induction is that it uses a "stronger" inductive hypothesis. The form of induction introduced in Theorem A.1 is sometimes called "weak induction" since it employs a "weaker" hypothesis, but we are more inclined to use the term "regular induction." In fact, it can be shown that the two forms of mathematical induction are equivalent; i.e., each implies the other.

Example A.4. We now give an example of the use of strong induction. We prove the existence part of the Fundamental Theorem of Arithmetic (see Theorem 2.4 for a detailed statement of this theorem). Let S be the set of positive integers greater than 1 that are either primes, or can be written as products of primes. Clearly $2 \in S$ since 2 is a prime. (Note that in this case we are not starting the induction from the case $k = 1$). We now *assume* that for some positive integer k, the set S contains all integers m with $2 \leq m \leq k$. We must show that $k+1$ is also in S, for then all m with $2 \leq m \leq k+1$ will be in S by our induction hypothesis.

If $k+1$ is a prime, then it is in S by the definition of S. Assume that $k+1$ is not a prime, so it can be written in the form $k+1 = m_1 m_2$ where $1 < m_1 < k+1$ and $1 < m_2 < k+1$. From the (strong) induction hypothesis, we know that m_1 and m_2 are in the set S. Thus each of the integers m_1 and m_2 is either a prime or is a product of primes. Hence when we multiply m_1 and m_2 together, the resulting product $k + 1$ will also be a product of primes and the proof is complete. We note that regular induction would not have worked here since simply knowing that k is in S would not imply that $k + 1$ is in S. Rather, we need to know about k and all m below k as well.

Having been given this brief introduction to the technique of mathematical induction, we now ask you to work through a number of other applications of this technique. We suggest that you try to solve the problems yourself before looking at the given solutions. Also, you may find it necessary to employ strong induction on occasion.

Solved Problems

A.1. Define a sequence of positive integers a_n be letting $a_1 = 1$ and then letting $a_{n+1} = 2a_n + 1$ for each $n \geq 2$. Show for each positive integer n that $a_n + 1$ is a power of 2.

Solution:
Use induction on the positive integer n. Since $a_1 = 1, a_1 + 1 = 2$, which is a power of 2. Just to see one more step above this base case, $a_2 = 2a_1 + 1 = 3$, so $a_2 + 1 = 4$, which again is a power of 2.

For the inductive step, we assume for $n = k$ that we have $a_k + 1 = 2^m$ for some positive integer m. Then

$$a_{k+1} + 1 = (2a_k + 1) + 1 = 2(a_k + 1) = 2(2^m) = 2^{m+1}.$$

A.2. For each positive integer n, show that $3^{2n} - 1$ is divisible by 8.

Solution:
The base case is $3^{2(1)} - 1 = 8$. Going one step further, the second case is $3^4 - 1 = 80$. For the inductive step, assume for $n = k$ that $3^{2k} - 1 = 8m$ for some positive integer m. Then

$$3^{2(k+1)} - 1 = 3^{2k}3^2 - 1 = (3^{2k} - 1)9 + 8 = (8m)9 + 8 = 8(9m + 1)$$

and we are finished.

A.3. Show that

$$1 \cdot 2 + 2 \cdot 3 + 3 \cdot 4 + \cdots + n(n+1) = \frac{n(n+1)(n+2)}{3}.$$

Solution:
Use induction on the positive integer n. For $n = 1$ we have $1(2) = 1(2)(3)/3$. Assume now that for $n = k$ we have

$$1(2) + 2(3) + \cdots + k(k+1) = \frac{k(k+1)(k+2)}{3}.$$

Then for $n = k + 1$ we have

$$1(2)+2(3)+\cdots+k(k+1)+(k+1)(k+2) = \frac{k(k+1)(k+2)}{3}+(k+1)(k+2),$$

which can be simplified to $\frac{(k+1)(k+2)(k+3)}{3}$, which completes the proof.

A.4. Find a simple closed formula (i.e., a formula with no summation in it) for the sum of the first n *even* positive integers.

Solution:
We can make use of Example A.1:

$$2+4+6+\cdots+2n = 2(1+2+3+\cdots+n) = 2(n(n+1)/2) = n(n+1).$$

So, for example, $2 + 4 + 6 + 8 + 10 = 30 = (5)(6)$.

Supplementary Problems

A.5. Prove that

$$1 + 4 + 9 + \cdots + n^2 = \frac{n(n+1)(2n+1)}{6}.$$

(Note: Some significant algebraic manipulation is needed here.)

A.6. For each positive integer n, show that $n^5 - n$ is divisible by 5. (Suggestion: In the induction step, use the Binomial Theorem to expand $(k+1)^5$.)

A.7. (a) Find a closed formula for the sum of the first n odd positive integers. (Suggestion: This sum is the difference $\sum_{k=1}^{2n} k - \sum_{k=1}^{n} 2k$. Now make use of Example A.1 and Problem A.4.)
(b) Using your answer to Part(a), do an induction proof to show that this answer is indeed correct.

A.8. Show, for every integer $n \geq 0$, that

$$2^0 + 2^1 + 2^2 + \cdots + 2^n = 2^{n+1} - 1.$$

A.9. The *Fibonacci numbers* F_n are defined by

$$F_1 = 1, F_2 = 1, \text{ and } F_{n+2} = F_n + F_{n+1};$$

i.e., each new Fibonacci number is the sum of the previous two.
(a) Write down the first 10 Fibonacci numbers.
(b) Use induction to prove that $\sum_{k=1}^{n} F_k = F_{n+2} - 1$.

A.10. (a) Write each of the numbers 30 and 35 as sums of *distinct* powers of 2.
(b) Use strong induction to prove that every positive integer n can be written a sum of distinct powers of 2. (Suggestion: For the inductive step, given n, suppose that largest power of 2 less than or equal to n is 2^k. Then $n = 2^k + m$ for some $0 \leq m < n$.)

Answers to Selected Supplementary Problems

A.7 (a) n^2.

A.9 (a) 1, 1, 2, 3, 5, 8, 13, 21, 34, 55.

A.10 (a) $30 = 16 + 8 + 4 + 2$, $35 = 32 + 2 + 1$.

Appendix B

Sets of Numbers Beyond the Integers

The integers $\mathbb{Z}$ are the object of study in number theory, but there are numerous points in this text where we refer to the rational numbers $\mathbb{Q}$, the real numbers $\mathbb{R}$, and the complex numbers $\mathbb{C}$. In particular, we have noted that these three number sets are *fields* (i.e., every non-zero element has a multiplicative inverse), whereas $\mathbb{Z}$ is definitely not a field since only 1 and -1 have multiplicative inverses. In this brief appendix, we discuss some basic properties of these sets and how they are related to each other.

As sets we have that $\mathbb{Z} \subset \mathbb{Q} \subset \mathbb{R} \subset \mathbb{C}$, where the symbol $\subset$ means "is a proper subset of." In diagram form we have

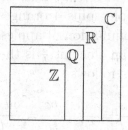

As algebraic objects, all four sets share the same operations of addition, subtraction and multiplication, and then the three larger sets share a division operation. As we move from one set to the

DOI: 10.1201/9781003193111-B

next larger one, those operations simply get applied to a wider collection of elements.

We shall now outline how, starting with the integers, we can move to the next larger set from the previous one.

Moving from the integers $\mathbb{Z}$ to the rational numbers $\mathbb{Q}$:

We start by adding to the integers the multiplicative inverses b^{-1} of every non-zero b in $\mathbb{Z}$. We can denote that number by $\frac{1}{b}$, so the product of $a \in \mathbb{Z}$ with b^{-1} is denoted by $\frac{a}{b}$. Hence we have the following simple description of the rational numbers:

$$\mathbb{Q} = \{\frac{a}{b} | a, b \in \mathbb{Z}, b \neq 0\}.$$

Multiplication in $\mathbb{Q}$ is given by $(\frac{a}{b})(\frac{c}{d}) = \frac{ac}{bd}$. Obviously the multiplicative inverse of $\frac{a}{b}$ is $\frac{b}{a}$ (provided that $a \neq 0$), and so our new operation of division in $\mathbb{Q}$ is just

$$\frac{\frac{c}{d}}{\frac{a}{b}} = (\frac{c}{d})(\frac{b}{a}) = \frac{bc}{ad}.$$

It is also clear that $\mathbb{Z}$ sits inside $\mathbb{Q}$ since $a \in \mathbb{Z}$ is simply $\frac{a}{1}$ in $\mathbb{Q}$.

Addition and subtraction in $\mathbb{Q}$ are a bit more complicated than multiplication and division because of the basic fact that to add (or subtract) two quantities, they must be "of the same denomination." What is 3 apples plus 4 oranges? Well, it's 7 pieces of fruit, the common denomination of apples and oranges. Likewise, the "denomination" of an element $\frac{a}{b}$ is the denominator b, and so addition in $\mathbb{Q}$ is done as follows;

$$\frac{a}{b} + \frac{c}{d} = \frac{ad}{bd} + \frac{bc}{bd} = \frac{ad + bc}{bd}.$$

We note that both addition and the above multiplication defined in $\mathbb{Q}$ are compatible with the corresponding operations in $\mathbb{Z}$ since $\frac{a}{1} + \frac{c}{1} = \frac{a+c}{1}$ and $(\frac{a}{1})(\frac{c}{1}) = \frac{ac}{1}$. Hence we have constructed a "field of fractions" $\mathbb{Q}$ which extends the non-field $\mathbb{Z}$ and whose algebraic operations are compatible with those of $\mathbb{Z}$.

Moving from the rational numbers $\mathbb{Q}$ to the real numbers $\mathbb{R}$:

An element $\frac{a}{b}$ in $\mathbb{Q}$ can be written in decimal form by simply doing long division (i.e., repeated application of the Division Algorithm, Theorem 1.2). So, for example, $\frac{1}{2} = 0.5$, $\frac{11}{8} = 1.375$, and $\frac{1}{9} = 0.111\ldots$, where "..." means that the 1's repeat endlessly. To describe this situation in general, we define the *repetend* of a rational number written in decimal form to be a fixed sequence of digits of fixed length which repeats endlessly. Hence in the three examples given above, the first two are called "terminating decimals" and the last has a repetend of 1. As another example, the rational number $\frac{1}{7}$ in decimal form is $0.142857142857\ldots$; that is, its repetend is 142857 (see Example B.1 below).

The following theorem characterizes the decimal forms of rational numbers:

Theorem B.1. *The decimal representation of every rational number either terminates or contains a repetend.*

Proof. Suppose we are computing the decimal representation of the rational number $\frac{m}{n}$ where m is less than n. (If $m > n$, we start by doing long division to arrive at a "mixed number" $a + \frac{m_0}{n}$ where a is the "whole part" and $\frac{m_0}{n}$ is a "proper fraction.") Now, in doing long division, the Division Algorithm guarantees that at each step the remainder upon division by n will be one of the n numbers 0 through $n - 1$. If at any point the remainder is 0, then the representation terminates. Otherwise, there are $n - 1$ possible remainder values. Suppose we perform n division steps; then by the so-called "Pigeon Hole Principle," at least one of the $n - 1$ possible remainder values must appear twice. But as soon as a given remainder value appears a second time, the sequence of following remainder values must also repeat, and thus we have a repetend. This completes the proof. $\square$

Example B.1. We show below the long division of the rational numbers $\frac{3}{8}$ and $\frac{1}{7}$. The former terminates, whereas the latter shows a repetend of length $6 = 7 - 1$, which is as long as possible. The reminders from the Division Algorithm are highlighted.

$$
\begin{array}{r}
0.\ 1\ 4\ 2\ 8\ 5\ 7 \\
7\,|\ \overline{1.\ 0\ 0\ 0\ 0\ 0\ 0} \\
7 \\
\hline
\mathbf{3}\ 0 \\
2\ 8 \\
\hline
\mathbf{2}\ 0 \\
1\ 4 \\
\hline
\mathbf{6}\ 0 \\
5\ 6 \\
\hline
\mathbf{4}\ 0 \\
3\ 5 \\
\hline
\mathbf{5}\ 0 \\
4\ 9 \\
\hline
\mathbf{1}
\end{array}
$$

$$
\begin{array}{r}
0.\ 3\ 7\ 5 \\
8\,|\ \overline{3.\ 0\ 0\ 0} \\
2\ 4 \\
\hline
\mathbf{6}\ 0 \\
5\ 6 \\
\hline
\mathbf{4}\ 0 \\
4\ 0 \\
\hline
\mathbf{0}
\end{array}
$$

We note that a terminating decimal could be described as having a repetend of 0 (for example, $\frac{1}{2}$ can be written $0.5000\ldots$), but we shall use the two categories ("terminating" and "having a repetend") as in the theorem. In fact, we ask you to discover which rational numbers are of one type and which of the other in Problem B.6.

We now come to the real numbers $\mathbb{R}$. Consider the decimal number:

$$s = 0.101001000100001\ldots.$$

There is a "pattern" here $(1, 01, 001,$ etc.$)$, but there is no repetend (remember that a repetend must be of a fixed length), and s does not terminate, so it cannot be a rational number by Theorem B.1. We make then the following definition:

Definition B.1. The set of *real numbers* $\mathbb{R}$ is the set of all possible decimal numbers; that is, the set of all decimal representations with a finite number of digits to the left of the decimal point and a finite or infinite number of digits to the right of the decimal point.

If a real number r is not a rational number, then r is called an *irrational number*. Hence the number s above is irrational. Two examples of "famous" irrational numbers are

$$\sqrt{2} = 1.4142135623730950488\ldots, \pi = 3.1415926535897932385\ldots.$$

We ask you to prove that $\sqrt{2}$ is irrational in Problem B.7. That π is irrational is harder to prove.

As an example of contrasting the sets $\mathbb{Q}$ and $\mathbb{R}$, let us consider the following question: Given a quadratic polynomial $f(x) = ax^2 + bx + c$ where a, b, c are integers and $a \neq 0$, under what conditions do the roots of $f(x)$ lie in $\mathbb{Q}$ or, if not, in $\mathbb{R}$? (Recall: By the *roots* of $f(x)$ we mean the solutions of the equation $f(x) = 0$.) The following result tells us the answer.

Theorem B.2. *Suppose that $f(x) = ax^2 + bx + c$ where a, b, c are integers and $a \neq 0$. Then*
(a) *the roots of $f(x)$ lie in $\mathbb{Q}$ if and only if $b^2 - 4ac$ is a perfect square, and*
(b) *the roots of $f(x)$ lie in $\mathbb{R}$ if and only if $b^2 - 4ac \geq 0$.*

Proof. Both of these assertions follow from the Quadratic Formula, which says that the roots of $f(x)$ are

$$x = \frac{-b \pm \sqrt{b^2 - 4ac}}{2a}.$$

For Part (b) we use the fact that in $\mathbb{R}$ we cannot take the square root of a negative number since any real number squared is non-negative. $\square$

Example B.2. (a) The roots of $2x^2 + 5x + 2$ are rational since $5^2 - 4 \cdot 2 \cdot 2 = 25 - 16 = 9$, which is a perfect square. In fact the roots are -2 and $-\frac{1}{2}$.

(b) The roots of $2x^2+5x+1$ are real but irrational since $5^2-4\cdot2\cdot1 = 17$, which is non-negative but not a perfect square. In fact, the roots are $\frac{-5\pm\sqrt{17}}{4}$.

(c) The roots of $2x^2 + 5x + 4$ are not real since $5^2 - 4\cdot2\cdot4 = -7$.

So what are we to do about finding of the roots of a polynomial like the one in Part (c) of Example B.2? The problem here was our inability to take the square root of a negative number.

Moving from the real numbers $\mathbb{R}$ to the complex numbers $\mathbb{C}$:

We know that $\sqrt{-1}$ is not a real number (since $x^2+1 = 0$ has no real solutions), but we now allow it to be a legitimate component of our algebra. We denote the $\sqrt{-1}$ by i, and we note that $i^2 = -1$. Hence i is not a real number, but its square *is* a real number. We can now define the complex numbers.

Definition B.2. The *complex numbers* $\mathbb{C}$ are the set of all numbers of the form $a+bi$ where a and b are real numbers. The number a is called the *real part* of the number, and bi is called the *imaginary part* of the number.

We note that $\mathbb{R}$ is a subset of $\mathbb{C}$ since in the imaginary part b can be 0.

Concerning algebraic structure, addition and subtract are simply done "component-wise;" that is

$$(a + bi) \pm (c + di) = (a \pm c) + (b \pm d)i.$$

Multiplication is simply multiplying two binomials, using the fact that $i^2 = -1$:

$$(a + bi)(c + di) = ac + adi + bci + bdi^2 = (ac - bd) + (ad + bc)i.$$

Division is a bit trickier. To accomplish it we need to make use of the *complex conjugate* of $c + di$, which is $c - di$. We note that the product of a complex number with its conjugate is a *non-negative real number*:

$$(c + di)(c - di) = c^2 - cdi + cdi - d^2i^2 = c^2 + d^2.$$

Note also that this product is 0 if and only if both c and d are 0. So, in doing division in $\mathbb{C}$, we multiply top and bottom by the conjugate of the denominator:

$$\frac{a+bi}{c+di} = \frac{(a+bi)(c-di)}{(c+di)(c-di)} = \frac{ac+bd}{c^2+d^2} + \frac{bc-ad}{c^2+d^2}i.$$

Returning now to the quadratic polynomial $f(x) = ax^2+bx+c$ (where a, b, c are integers and $a \neq 0$) considered in Theorem B.2, we now see that $f(x)$ has two roots regardless of the value of $b^2 - 4ac$ *if we allow complex roots*. Going back to Example B.2 Part (c), we now have that the polynomial $2x^2 + 5x + 4$ has two roots, specifically $\frac{-5\pm i\sqrt{7}}{4}$.

We can, it turns out, greatly generalize this fact about quadratic equations: We can allow the coefficients of the given polynomial to lie in any of the sets $\mathbb{Z}$, $\mathbb{Q}$, $\mathbb{R}$, and/or $\mathbb{C}$, and the degree n of the polynomial can be any positive number, not just $n = 2$. Before stating this very important result we need to make a couple of points. First, when we say that a polynomial of degree n has n roots, it may be that one (or more) of them is appearing more than once (this is referred to as *multiplicity*). For example, the polynomial $f(x) = x^2 - 2x + 1 = (x-1)^2$ has the root 1 appearing twice, so we say 1 has multiplicity 2. Likewise, in $g(x) = (x-5)^3$, the root 5 has multiplicity 3, and so on. Second, when we say the polynomial has "complex coefficients" or "complex roots," remember that those coefficients and roots can be integers, rational numbers, or real numbers. Saying they are "complex" covers all those possibilities.

Theorem B.3. (The Fundamental Theorem of Algebra) *Every polynomial of positive degree n with complex coefficients has exactly n roots (counting multiplicities) in the complex numbers $\mathbb{C}$.*

So, if we are faced with finding the roots of a polynomial of degree 23, we may or may not be able to actually find them, but we do know that in the complex numbers, all 23 roots exist. One way to express this rather surprising property of $\mathbb{C}$ is that it is "algebraically closed." Note, for example, as we have seen in Theorem B.2, that this is definitely *not* a property of the rational or real numbers.

Finally, we wish to point out that in various places in this text, especially in Chapters 9 and 11, we have seen that in order to try to solve number theory problems (i.e., problems involving the integers $\mathbb{Z}$) it can be helpful, even necessary, to view $\mathbb{Z}$ as lying inside the real numbers $\mathbb{R}$ and/or the complex numbers $\mathbb{C}$. For two very important examples, Fermat's Last Theorem (Conjecture 9.6) and the Prime Number Theorem (Theorems 11.2 and Theorem 11.3) were both solved making very heavy use of the complex numbers. Analytic number theory, that is, using the structure of the real number $\mathbb{R}$ or complex numbers $\mathbb{C}$ to learn about the structure of their integer subset $\mathbb{Z}$, has turned out to be a very powerful technique. Hence to better understand $\mathbb{Z}$, we need some knowledge of $\mathbb{Q}$, $\mathbb{R}$, and $\mathbb{C}$, and this is what we've tried to supply in this appendix.

Solved Problems

B.1. Concerning addition in $\mathbb{Q}$, the given formula

$$\frac{a}{b} + \frac{c}{d} = \frac{ad}{bd} + \frac{bc}{bd} = \frac{ad + bc}{bd}$$

may not give the answer in "lowest terms." (We say that $\frac{a}{b}$ is in *lowest terms* if $\gcd(a, b) = 1$.) Write down a formula for addition which results in an answer which is in lowest terms and give an example to illustrate your formula.

Solution:
Given $\frac{a}{b} + \frac{c}{d}$, we want the denominator of the answer to be the least common multiple of b and d, which is $bd/\gcd(b, d)$. If we denote $\gcd(b, d)$ by g, then dividing top and bottom of our addition

formula by g, we get

$$\frac{a}{b} + \frac{c}{d} = \frac{ad + bc}{bd} = \frac{a(\frac{d}{g}) + (\frac{b}{g})c}{\frac{bd}{g}}.$$

For example, since $\gcd(6, 8) = 2$, we get

$$\frac{5}{6} + \frac{3}{8} = \frac{5(\frac{8}{2}) + (\frac{6}{2})3}{\frac{6 \cdot 8}{2}} = \frac{29}{24}.$$

B.2. We know that we can use long division to convert a rational number from fractional form to decimal form. Show how you can go the other direction by manipulating the decimals (a) 0.143 (b) 0.143143143... (c) 0.6131313....

Solution:
(a) Since the decimal terminates, we can simply divide by the appropriate power of 10 to move the decimal to the right. Since 0.143 has three decimal places, we divide by $10^3 = 1000$. Hence the answer is $\frac{143}{1000}$, which is already in lowest terms.
(b) Let $x = 0.143143....$ Multiply both sides by 1000: $1000x = 143.143143....$ Now subtract the first equation from the second: $999x = 143$, so we get $x = \frac{143}{999}$.
(c) Let $x = 0.61313....$ First multiply both sides by 10 to move the 6 out of the decimal part: $10x = 6.1313....$ Now since the repetend is of length 2, multiply both sides of the second equation by 100: $1000x = 613.1313....$ Finally subtract these latter two equations, obtaining $990x = 607$, and so $x = \frac{607}{990}$.

B.3. Prove the Quadratic Formula, i.e., prove that if a, b, c are from any of the sets $\mathbb{Z}$, $\mathbb{Q}$, $\mathbb{R}$ or $\mathbb{C}$, then the roots of $p(x) = ax^2 + bx + c$ are

$$x = \frac{-b \pm \sqrt{b^2 - 4ac}}{2a}.$$

Solution:
We use the technique of "completing the square:"

1. Starting with the equation $ax^2 + bx + c = 0$, divide through by a and subtract $\frac{c}{a}$ from both sides:

$$x^2 + \frac{b}{a}x = -\frac{c}{a}.$$

2. Complete the square by adding $(\frac{b}{2a})^2$ to both sides:

$$x^2 + \frac{b}{a}x + (\frac{b}{2a})^2 = (\frac{b}{2a})^2 - \frac{c}{a}.$$

3. Factor the left-hand side and simplify the right-hand side:

$$(x + \frac{b}{2a})^2 = \frac{b^2 - 4ac}{4a^2}.$$

4. Take the square root of both sides, subtract $\frac{b}{2a}$ from both sides, and combine the right-hand side into a single fraction, obtaining

$$x = \frac{-b \pm \sqrt{b^2 - 4ac}}{2a}.$$

B.4. Find all 5 roots of the polynomial $x^5 + x^4 + x^3 + x^2 - 6x - 6$.

Solution:
It is not easy to find the roots of a 5-th degree polynomial; you need some luck. One possibility is "guess and check;" i.e., plug in a few small numbers and see if you get an answer of 0. This happens to work in this case: -1 is a root. Hence our polynomial is divisible by $x + 1$, and upon division the remaining factor is $x^4 + x^2 - 6$. But this is a polynomial in x^2 which can be factored into $(x^2 - 2)(x^2 + 3)$, so we see that our other four roots are $\sqrt{2}$, $-\sqrt{2}$, $i\sqrt{3}$, and $-i\sqrt{3}$. Thus our polynomial has one integer root, two irrational roots, and two complex roots.

Supplementary Problems

B.5. (a) By converting $\frac{5}{13}$ to a decimal, find its repetend.
(b) Convert $3.24444\ldots$ (i.e., the repetend is 4) to a fraction in lowest terms.

B.6. (a) Convert $\frac{1}{2}, \frac{1}{3}, \ldots, \frac{1}{10}$ to decimal form.
(b) Looking at this small amount of data, given a denominator n, make a conjecture as to when $\frac{1}{n}$ terminates and when $\frac{1}{n}$ has a non-trivial repetend.
(c) Prove your conjecture.

B.7. Prove that $\sqrt{2}$ is irrational. (Suggestion: Suppose not; i.e., suppose that $\sqrt{2} = \frac{a}{b}$, a rational number in lowest terms. Multiply through by b, square both sides, and then look closely at the resulting integer equation.)

B.8. Find the three roots of $3x^3 - 13x^2 + 19x - 10$. (Suggestion: Try plugging in a few small positive integers a. If the answer is 0, you've found a root, and after dividing by $x - a$, you'll have a quadratic to work with.)

Answers to Selected Supplementary Problems

B.5. (a) 384615, (b) $\frac{292}{90} = \frac{146}{45}$.

B.8. $\{2, \frac{7+i\sqrt{11}}{6}, \frac{7-i\sqrt{11}}{6}\}$.

Bibliography

[1] G.E. Andrews, *Number Theory*, Saunders, Philadelphia, 1971; Reissued by Dover, 1995.

[2] W. Diffie and M.E. Hellman, New directions in cryptography, *IEEE Transactions on Information Theory* IT-22 (1976), 644-654.

[3] E. Grosswald, *Topics from the Theory of Numbers*, 2nd Ed., Birkhäuser, Boston, 1984.

[4] C.F. Laywine and G.L. Mullen, *Discrete Mathematics Using Latin Squares*, Wiley-Interscience Series in Discrete Mathematics and Optimization, John Wiley and Sons, New York, 1998.

[5] R. Lidl and H. Niederreiter, *Introduction to Finite Fields and Their Applications*, Revised Ed., Cambridge University Press, Cambridge, 1994.

[6] R. Lidl and H. Niederreiter, *Finite Fields*, Cambridge University Press, *Encyclo. Math and Appls.*, Vol. 20, Sec. Ed., Cambridge University Press, 1997.

[7] G.L. Mullen and C. Mummert, *Finite Fields and Applications*, Amer. Math. Soc., Student Mathematical Library, Vol. 41, Providence, RI, 2007.

[8] G.L. Mullen and D. Panario, *Handbook of Finite Fields*, CRC Press, Taylor and Francis Group, Boca Raton, FL, 2013.

[9] G.L. Mullen and D. White, A polynomial representation for logarithms in $GF(q)$, *Acta Arithmetica* 47(1986), 255-261.

[10] I. Niven and H.S. Zuckerman, *An Introduction to the Theory of Numbers*, 4th Ed., Wiley, New York, 1980.

[11] R.I. Rivest, A. Shamir, and L. Adleman, A method for obtaining digital signatures and public-key cryptosystems, *Comm. ACM* 21(1978), 120-126.

[12] J.H. Silverman, *A Friendly Introduction to Number Theory*, 4th Ed., Pearson Prentice Hall, Upper Saddle River, NJ, 2013.

Index

Printed in the United States
by Baker & Taylor Publisher Services